Sins against Science

SUNY series, Studies in Scientific and Technical Communication
James P. Zappan, editor

Sins against Science

The Scientific Media Hoaxes of Poe, Twain, and Others

Lynda Walsh

State University of New York Press

Cover: Lunar Discoveries, image with caption from *New York Sun*, Oct. 16, 1835, p. 4, negative number 77957d, Collection of the New-York Historical Society.

Published by
State University of New York Press, Albany

Printed in the United States of America

For information, address State University of New York Press, 194 Washington Avenue, Suite 305, Albany, NY 12210-2384

Production by Marilyn P. Semerad
Marketing by Fran Keneston

Library of Congress Cataloging-in Publication Data

Walsh, Lynda, 1971–
Sins against science : the scientific media hoaxes of Poe, Twain, and others / Lynda Walsh.
p. cm. — (SUNY series, Studies in scientific and technical communication)
Includes bibliographical references and index.
ISBN-13: 978-0-7914-6877-7 (hardcover : alk. paper)
ISBN-10: 0-7914-6877-1 (hardcover : alk. paper)
ISBN-13: 978-0-7914-6878-4 (pbk : alk. paper)
1. Scientific literature. 2. Science news. 3. Impostors and imposture in literature. I. Title. II. Series.

Q225.5.W35 2006
500—dc22
2005033336

10 9 8 7 6 5 4 3 2 1

Contents

List of Illustrations vii

Acknowledgments xi

Introduction 1
- Previous Work 2
- Relevance of Hoaxes to Current Scholarly Concerns 4
- Chapter Summaries 15

Chapter One: A Brief Natural History of Hoaxing 17
- Swift's Hoax and Satires 18
- Parody 21
- Nineteenth-Century Fraud, Tall Tales, and Science Fiction in America 23
- Kairos 27

Chapter Two: Method 35
- The Traditional Genealogy 35
- The Subversive Genealogy 39
- Optimality Theory 42

Chapter Three: Poe's Hoaxing and the Construction of Readerships 51
- Overview of Poe's Scientific and Rhetorical Acculturation 52
- The Contest between Hans Phaall and Locke's Moon Hoax: Revealing Reader Expectations 60
- Collecting Reader Expectations 66
- Ranking Reader Expectations 85
- The Balloon-Hoax 90
- "The Facts in the Case of M. Valdemar" 97
- "Von Kempelen and His Discovery" 102
- Solutions to Problems in Poe Scholarship 107

Poe's Relationship to Science and to His Readership: How the Hoaxes Interact with *Eureka* 114

Chapter Four: Mark Twain and the Social Mechanics of Laughter 121
Rhetorical Acculturation 122
Scientific Acculturation 127
The Petrified Man 132
Adjusting the Filter of Expectations to Account for Twain's Hoaxing 154
Applying the Analysis to Problems in Twain Scholarship 157
Relationship of the Hoax to Twain's Scientific Thinking 161
The Social Mechanics of Laughter 166

Chapter Five: The Hoaxes of Dan De Quille: Building and Defending the West 173
Rhetorical Acculturation 173
Scientific Acculturation 176
De Quille's Hoaxes 179
Summary of Reading Expectations Based on De Quille's Hoaxes 204
De Quille's Hoaxes Build and Defend His Ideal West 206

Chapter Six: The Mechanics of Hoaxing 213
How Did the Hoaxes Work? 214
What Were the Hoaxers Trying to Accomplish? 215
The Hoax as a Machine 220
Tuning Up: The Hoax Then and Now 224

Conclusion: The Sokal Hoax 227
Exploiting the Conventions of the Cultural Studies Article 228
Sokal's Hoax Constructs Him as a Notorious Expert 232
The Hoax as a Computer Virus 235

Appendix A: How to Read Tables in Optimality Theory (OT) 241
What the Parts of the Table Mean 241
The Results of the Syllabification of /ansɛt/ 242
Optimality Theory Applied to a Decision about a Hoax's Truth-Value 242

Appendix B: Reader Responses to the Moon Hoax 245

Notes 253

Glossary 275

Bibliography 279

Index 291

Illustrations

Figures

I.1.	Graphical Representation of Reader Expectations: Moon Hoax	13
I.2.	An Optimality Theory Model of the Reader's Decision in Figure I.1	14
3.1.	Richard Adams Locke, author of the Moon Hoax, Courtesy, Library of Congress	64
3.2.	J. W. F. Herschel by Julia Cameron, 1858, Courtesy, Wilson Centre for Photography	65
3.3.	The First Page of Poe's Balloon Hoax. Courtesy, American Antiquarian Society	96
5.1.	Dan De Quille (William Wright), Courtesy, Special Collections, University of Nevada-Reno Library	181
5.2.	Nevada Miners, Courtesy, Special Collections, University of Nevada-Reno Library	182
5.3.	Composing Room at Territorial Enterprise Showing Desk used by De Quille and Twain, Courtesy, Special Collections, University of Nevada-Reno Library	184
5.4.	Virginia City, Nevada. Photo of Lithograph, Courtesy, Special Collections, University of Nevada-Reno Library	185

Tables

2.1. Syllabification of "onset"/ansɛt/ 43

2.2. Decision about Tsun's Narration of Children 46

3.1. Expectations of Popular Science News Collected from Reader Responses to Moon Hoax 69–71

3.2. Salient Features of "Moon Hoax" Parodied by *Herald* 72

3.3. Antebellum Media Surveyed with Number of Science Articles per Issue 74

3.4. Distribution of Categories of Science Articles across Media Sample 75

3.5. Typical Opening of Antebellum Science Articles 77

3.6. Typical Structuring of "Problem" Phase of Antebellum Science Articles 78

3.7. Typical Structuring of "Solution" Phase of Antebellum Science Articles 79

3.8. Poe's Characteristics of a Successful Hoax 82

3.9. Reader Michael Floy's Decision to Disbelieve Locke's Moon Story 86

3.10. NY *Commercial Advertiser* Editor's Decision to Disbelieve Locke's Moon Story 87

3.11. Graph of Decision by Poe's Projected Readers to Believe in Locke's Moon Bison 88

3.12. Decision about the Truth-Value of Poe's Balloon Hoax by Readers Valuing Plausibility and the Reputation of the News Medium 92

3.13. Suspended Decision about the Truth-Value of Poe's Facts in the Case of M. Valdemar by E. B. Barrett 99

3.14. Suspended Decision about the Truth-Value of Poe's Facts in the Case of M. Valdemar by A. Ramsay 100

4.1. Decision of Rival Editors about The Petrified Man 134

4.2. Editors' Projection of their Readers' Interpretive Process 136

4.3. Skimming Guided by Newsreading Conventions 146

4.4. 1865 Media Surveyed with Number of Science Articles per Issue 148

4.5. Distribution of Categories of Science Articles across 1865 Media Sample 149

4.6. Opening Structure of 1865 Popular Science Articles 151

4.7. Structure of "Problem" Phase of 1865 Popular Science Articles 152

4.8. Structure of "Solution" Phase of 1865 Popular Science Articles 153

5.1. Reader Reaction to Silver Man Projected by De Quille's *Refutatio* 183

5.2. Editor's Suspended Judgment about Solar Armor Hoax 192

6.1. Comparison of Filters of Reading Expectations from the Project 218

A.1. Syllabification of /ansɛt/ in OT 241

A.2. A Reader's Decision to Believe the Moon Hoax Based on Spectacle, Not Science 243

Acknowledgments

I would like to thank the following people who have contributed to this book: Davida Charney for her guidance and mentoring—I could not ask for a better advisor and role model; my husband, Patrick, for his advice, support, and encouragement; David Kessler and the staff at the reading room of the Bancroft Library at Berkeley; Jacque Sundstrand in the Special Collections Department of the University of Nevada–Reno Library; the staff at the New-York Historical Society reading room; Marie Secor for her collaboration on an alternate version of the Sokal analysis; Sharan Daniel for her friendship, experience, and sharp intuition; the anonymous reviewers of this project for SUNY Press who provided gracious and invaluable criticism; and my family and friends for being brave enough to keep asking how the book was coming along.

I would like to thank the following institutions for the permission to reprint previously published sections of this work:

- Parts of chapter 2 appeared as "The Scientific Media Hoax: A Rhetoric for Reconciling Linguistics and Literary Criticism," in *Rhetorical Agendas: Political, Ethical, Spiritual*, ed. Patricia Bizzell (Mahwah, NJ: Lawrence Erlbaum and Associates, 2006), 165–78.
- Parts of chapter 3 appeared as "What Is a Hoax? Redefining Poe's Jeux d'Esprit and His Relationship to His Readership," in *Text, Practice, Performance* 4 (2002). Available online at http://www.utexas.edu/cola/depts/culturalstudies/tpp/archive.html.
- Parts of chapter 5 appeared as "Dan De Quille and the Scientific Hoax as a Technology of Resistance," in *Popular Culture Review* 15(1), 39–50.

- Parts of the conclusion were previously published as Marie Secor and Lynda Walsh, "A Rhetorical Perspective on the Sokal Hoax: Genre, Style, and Context," *Written Communication* 21(4) pp. 69–91, copyright 2004 by Marie Secor and Lynda Walsh. Reprinted by permission of Sage Publications, Inc.

And finally, I return this work to Him from Whom I have received it: Jeremiah 33:3.

Introduction

In October of 1862, the *Territorial Enterprise* of Virginia City, Nevada, reported the startling discovery of a fully intact fossilized man. The author and city editor of the paper, Mark Twain, carefully reported all the scientific particulars of the petrifaction to a readership eager for more of the recent revelations of geologic wonders such as ice ages and fossilized mammoths. Twain's story was immediately picked up by eastern and western newspapers and was wired perhaps as far as London—all in spite of the fact that a careful reading of the narrative betrays that the man was fossilized sitting up and thumbing his nose at the reader.[1]

Twain was not the only writer to toy with America's fascination with science in the nineteenth century. From 1835 to 1880, at least a dozen similar hoax science stories appeared in penny dailies or literary monthlies. Edgar Allan Poe wrote four of them. Other contributors included Richard Adams Locke, lateral descendant of John Locke and perpetrator of the famous Moon Hoax of 1835 that sold a record number of copies of the New York *Sun*; and Dan De Quille, a coeditor of Twain on the *Enterprise*, who authored numerous scientific hoaxes including the Solar Armor hoax in which an inventor turns up a frozen corpse in the middle of Death Valley because the wet-sponge suit he designed to cool him down worked a little too efficiently.[2]

This roughly thirty-year heyday of scientific media hoaxing remains an unexamined and unexplained phenomenon in the history of American literature and popular science. I discovered it by accident through a stray line mentioning it in Robert V. Bruce's *Launching of Modern American Science*. I was fascinated and instantly had two questions I wanted answered about the hoaxes: Why did august literary figures such as Poe and Twain stoop to writing fake science stories for newspapers? And how had they managed to fool so many people? I reviewed the scholarly work on these hoaxes, but my questions remained unanswered. What previous scholars had focused on was trying to reconcile the hoaxes of each author with the rest of his literary canon. What I wanted to do was to bring into view all of the hoaxes at once and consider them as a cultural event—a response by the American literary community to the ascendance of science as a social epistemology in the mid-nineteenth century. In other words, my questions were not answered because they were not literary questions but rather questions of pragmatics—the work words do in the world. I wanted to know how the hoaxes had managed to change, however briefly, the world views of their readers. I also wanted to see what the hoaxes could tell me about the negotiation of scientific truth as a public commodity in America.

Previous Work

The previous work on hoaxing that I encountered was hampered overall by a New-Critical fixation with the physical text that was ill-equipped to cope with the crucial component of the hoaxes' reception as well as their tendency to shift generic categories over time. These treatments could neither distinguish hoaxes from related genres nor describe the unique relationships a hoax constructs among author, audience, and medium. So, while they were very useful in establishing the historical, factual contexts of the hoaxing events, I knew I would have to look beyond them to answer my research questions.

Most popular and scholarly coverage of media hoaxes remains essentially anecdotal. By far the most thorough is Alexander Boese's *Museum of Hoaxes*, which although it presents itself as a historical catalog of hoaxes, still includes a trenchant introductory analysis of hoaxing as a cultural phenomenon based on Boese's doctoral work in science studies at the University of California, San Diego. Before Boese's recent contribution, Fred Fedler's historical survey, *Media Hoaxes*, provided all the publication facts of many of the hoaxes considered in this book along with plot summaries and stories of their effects on readers. However, Fedler offered no analysis of why the hoaxes worked or did not, and he dismissed the phenomenon as a side effect of the lack of objective standards plaguing nineteenth-century news.

Americanists working on Poe and Twain, especially, have had to confront these authors' hoaxing practices at some point, and this literature will be reviewed more completely in the chapters on the individual authors. However, in general I found these treatments to assume rather than define what they meant by the word *hoax* and to ignore some crucial social complexities of hoaxes—especially their invocation of a double audience of readers who "fall for it" and readers who "get it"—in their eagerness to make the hoax work as a trope for Poe's and Twain's disillusionment with industrial society.

I found the most fruitful studies of hoaxes, counterfeits, and fakes outside of literary studies in the fields of social history and science and technology studies. Hugh Kenner's landmark study of the social function of fakery, *The Counterfeiters*, argues that the principal social benefit of a counterfeit is the hyperawareness it confers upon its viewers, once they have recognized its artifice, of the "realness" of the object or skill the counterfeit is imitating.[3] So in Kenner's scheme, a hoax such as Maelzel's chess-playing automaton, once Edgar Allan Poe exposed it as a fake, served mainly to highlight the inimitable power of the human strategic faculty. Viewers who had originally marveled at the power of the machine now sniffed: well, of course a machine could not play chess like a man! This special codependence between artifice and the real thing, between belief and doubt, will prove crucial to my own approach to hoaxes, which seeks to locate their cultural work in the changes they facilitate to readers' world views.

Why has there been no systematic scholarly treatment to date of this burst of scientific media hoaxing from 1835 to 1880? First, media hoaxes in general have suffered from their association with popular culture, which at least until the 1960s was considered beneath the dignity of literary study. Certainly, communications and journalism departments have been consistently concerned with the historical rhetoric of the news, but in these types of analyses, of which Fred Fedler's is a prime example, hoaxes are dismissed as a funny but unfortunate epiphenomenon of the "Dark Ages" of the penny press and yellow journalism. More problematic for the study of hoaxing, however, is an active folk definition of the practice that contains promising elements but that fools us into thinking we know what hoaxing is without having to explain or examine our assumptions. All of the studies reviewed above, excepting Boese's, labor under this folk definition; they do not offer working definitions of hoaxing because they assume we all know what the newscaster means when she reports on the latest "anthrax hoax," the "Sokal hoax," or a "fossil hoax." In daily use, 'hoax' overgenerates to apply to any situation in which the public initially mistakes an object or communication. It also connotes a sense that someone has intended this misapprehension. This may be another cause for the critical neglect of hoaxes—their sticky association with the intentional fallacy. If people misapprehend a message or object, does the originator of the message or object have to have intended to fool people for it to be considered a hoax?

This project, then, seeks to answer the questions already mentioned—How did these nineteenth-century scientific media hoaxes fool their readers? And what were the hoaxers trying to accomplish with their hoaxes?—along with a third question that I found unavoidable as I began my research—What is a hoax? In order to answer these essentially pragmatic questions, I treat the hoaxes under examination as interactive social events. I frame them as reactions to and perpetuations of a particular kairos—an opportunity to speak up—prompted by increasing tensions between scientific and artistic cultures in antebellum America. Although each hoaxer had his own reasons for perpetrating his hoaxes, they all shared a common mechanics of illusion and revelation that served to foreground assumptions that newsreaders were making about science and science news; these assumptions were then open for criticism by the hoaxer and others. In addition, each hoax created for its author a sort of expert notoriety that extended a platform for launching further social criticism. This new rhetorical definition of hoaxing as a complex activity played out over a historical timeline accounts for its rich textuality, as the hoax itself lives outside the physical text as tense relationships among reader reactions, media reputation, and authorial intentions.

Relevance of Hoaxes to Current Scholarly Concerns

Because hoaxes operate at the nexus of scientific and literary epistemologies, because they adopt the rhetoric of popular media to criticize the specialized rhetorics of groups viewed as politically threatening, studying them necessitates multiple critical approaches. I will use a bricolage of techniques found within the disciplines of rhetoric of science, reader-response criticism, new-historicist criticism, and linguistic text pragmatics to answer these basic questions: how did hoaxers fool their readers, and to what ends? In return, the hoaxes contribute to the illumination of an issue that concerns all these schools of textual criticism: the negotiation of social reality between writers and readers through the mass media, in order to satisfy as many of the public's (potentially competing) desires and expectations as possible.[4]

The Rhetoric of Science and Its Popularization

Overall, hoaxes offer opportunities to scholars of historical scientific rhetoric in three areas. First, hoaxes are an intervention in the formation of "ethnoscience," or pervasive lay attitudes toward science; we can use the resistance that hoaxes pose to the process of popularization to structure that concept in more detail. Second, examining these hoaxes permits us to define two understudied genres: the scientific media hoax itself and the antebellum popular science article that it apes. As genres are rituals encoding common communicative transactions between writers and readers through media, understanding the conventions of the nineteenth-century popular science article and its hoaxes will help us understand the public reaction to science in this key period when it was gaining enough recognition to influence ethics and politics in America. Third, the method of this project is a response to recent calls by rhetoricians of science and technology for increased rigor in reconstructing the reception of science texts.

Science Popularization in Nineteenth-Century America

Recent discussions of popularization have focused on reframing the discipline at its most foundational level, asking if there really is any such thing as popularization, and if there is, what methods will elucidate and recuperate it historically? Ludwik Fleck's groundbreaking analysis of the propagation of scientific knowledge in *Genesis and Development of a Scientific Fact*, published in 1935, marks the functional beginning of the study of science popularization. Fleck argued for a feedback loop of support and authorization between esoteric scientific communities and "exoteric" publics that revised traditional assumptions that scientific popularization was a one-way street from the lab to the lay public. Roger Cooter and Stephen Pumfrey have recently argued that key studies in the development of popularization studies have forced revisions of even Fleck's two-way model. Chief among them are Michel Callon and John Law's extension of Bruno Latour's study of laboratory ecologies, which shows the enrollment of private and government agencies to be a third factor in the success of scientific research; in addition, studies of lay artisan communities such as Adrian Desmond's study of nineteenth-century British craftsmen have revealed that communities of nonspecialists have consistently created their own scientific knowledge from praxis and through revision of the claims of "specialized" science to fit their own agendas.[5] These complications to the binary science/public models of popularization have cast doubt on the very definition of the field of study. Is there really any such phenomenon as "popularization," or is that simply a convenient reduction of the enormous, complicated work of making science to a more manageable dynamic that runs from lab to street? There is no simple solution to this problem, but Cooter and Pumfrey call for more historical studies of the making of public scientific knowledge to help put the study of science popularization into proper context.[6]

Cooter and Pumfrey's call has been answered by scholars working on the twentieth century and on nineteenth-century Britain, most of whom have concluded that science in those periods was sold to the lay public as technology or paraded as spectacle: Dorothy Nelkin studied press coverage of science since World War I in *Selling Science*; Jeanne Fahnestock and Greg Myers have both inquired specifically into the adjusting of scientific claims to public desire, finding important shifts in occasions and topoi; Charles Bazerman found that contributors to the *Philosophical Transactions of the Royal Society* cast their scientific work as revealing the spectacle of God's creation for "novelty-hungry" readers;[7] Steven Shapin studied the use of methods descriptions and engravings by Robert Boyle to create a virtual experience of his work and thus extend its credibility by multiplying witnesses.

Studies similar to these have helped structure the "wonder-business," as Mark Twain deemed it, of bringing science news to the public—especially in the seventeenth, eighteenth, and twentieth centuries.[8] Recently, researchers such as Jan Golinski, James Secord, and Greg Myers have extended the inquiry to nineteenth-century Britain, using reception data to identify key topoi that shaped public response to scientific activity at specific junctures in the social history of science.

These careful methods need now to be applied to nineteenth-century America.[9] Just as the industrial revolution in England was a pivotal period for the rhetoric of science, as laypeople began to talk and write about science and technology as major figures in their social landscape for the first time in history, so with the midnineteenth century in America. The scientific media hoaxes identified for study in this project constitute an intervention in the process of science popularization during this time. Hoaxers such as Poe and Twain exploited reader assumptions about science and science news to fool their readers into believing the hoaxes were true witnesses to scientific reality. Then, by revealing the hoaxes, either textually or contextually, the authors exposed the unconscious expectations the public had about science as vehicle to the Truth and implied that those expectations were unwarranted.

Poe and Twain could only accomplish this effect by successfully identifying and reperforming in their hoaxes their readers' expectations about science and science writing. These common expectations that readers held, like the topoi Myers identified in his study of the reception of thermodynamics in Victorian Britain, formed the units of exchange that science writers and readers traded in popular science articles and in the science hoaxes. Identifying these expectations, suggesting their origins, and showing how they might have interacted with each other in the hoax-reading process are major goals of this project and its principal contribution to the study of popularization. In the hoaxes we witness a unique game in metapopularization, as Poe and the other hoaxers strove to make the public aware of their being conditioned to accept scientific truth over the truths of faith, reason, and the imagination. Moreover, though the particular crisis between science and art that triggered these hoaxes in antebellum America is past, hoaxing is still alive and well as a strategy of intervention in the process of acculturating readers to either scientific or humanistic epistemologies. The conclusion to this book will extend findings from the historical section of the study to account for the controversy surrounding Alan Sokal's 1996 hoax in *Social Text*.

Scientific and Technical Genres

Scholars of scientific and technical genres are interested in how the genres are developed and codified over time and in how that process indexes the changing values and goals both of the scientist/engineer writers and of their readers.[10] Writing these kinds of histories of genres presents special problems of data collection and methodology that have recently elicited innovative solutions. David Kaufer and Kathleen Carley combined rhetorical analysis with quantitative models of knowledge dissemination over time in their study of print culture to argue that the development of scientific journals facilitated the flow of knowledge between disciplines starting in the eighteenth century.[11] Carol Berkenkotter and Thomas Huckin used read-aloud protocols of scientists reading research reports, coupled with textual surveys of 350 articles from 12 scientific journals over a period of 45 years, to show that the scientific article is still changing to reflect reader preferences in the late twentieth century. By contrast, Carolyn R. Miller focused her dissertation on a self-destructing technical genre. She analyzed the environmental impact statement (EIS) over time in concert with environmental legislation to demonstrate that the EIS failed to address competing interests in its governmental readership. These studies all combine diverse textual and contextual approaches to write histories of the feedback loop between genre formation and reader goals and values.

My study continues that line of inquiry through two genres—the scientific media hoax and the popular science article—over two periods in the nineteenth century—1835–1849 and 1862–1880. Like parodies and burlesques, hoaxes must mimic another genre for their effect—science news articles, in this case. In the case of the successful hoaxes, hoaxes that fooled thousands of readers, we know that they convincingly reperformed the salient features of the science news article. If imitation is the sincerest form of flattery—or perhaps the insincerest, in the case of hoaxing—it is also an excellent heuristic for reconstructing the rhetorical form of the mimicked genre. Much as John Swales has done for the academic article, this project hopes to structure the nascent popular science article, positioned crucially as it was at the juncture between specialized and popular communities of science and literature in the midnineteenth century. Since, as Dwight Atkinson argues in his study of the genre of the research article, key values and contemporary issues referenced in a scientific genre provide an index to "insider perspectives, and social practices or ideologies" of a given readership, the hoaxes can provide a suggestive index to the role science played in the lives of lay readers from 1835 to 1880.[12]

Methodological Problems in the Rhetoric of Science

A lingering problem in scientific and technical genre research, in spite of advances in historiographical and empirical methodologies, is a tendency to read the success of certain movements in science backwards into its rhetoric. In "Moving beyond the Moment," Danette Paul, Davida Charney, and Aimee Kendall argue that a narrow focus on the production of scientific texts often exaggerates their importance for their communities of reception. Instead, the authors argue, scholars of rhetoric of science should combine their analysis of the rhetoric of a text with analyses of its immediate reception and its changing reception over time. Only in this way can researchers avoid applying post hoc judgments, based on the eventual success or failure of the theories the author espoused, to the success or failure of the author's strategies for adapting his/her claims to readers' values and interests. A second, related methodological challenge is issued by Jeanne Fahnestock in "Rhetoric of Science: Enriching the Discipline." She calls for a more rigorous and consistent stylistics that enables the clear definition of problems in rhetorical analysis of scientific texts while simultaneously creating a system of accountability for results drawn from those analyses.

With my approach to the nascent genre of scientific hoaxing, I hope to answer the challenges issued by Paul, Charney, and Kendall and Fahnestock. First, I take as my primary object of study the immediate reception of scientific media hoaxes by four different authors living in different communities within the United States over a roughly thirty-year period. The immediate reception of the hoaxes is accessible via the contemporary media, private diaries and memoirs of readers, and commentary by the hoaxers themselves. Each hoaxer modified the genre of the hoax for his own purposes and to fit the cultural milieu in which he wrote, and a diachronic and geographic survey of all of these hoaxes will provide a detailed history, sensitive to time and place, of the development of a unique and unstudied scientific genre. Second, I structure my survey via a formal linguistic-pragmatic model, optimality theory. Working within the constraints of this theory and its related forms in cognitive psychology proffers several benefits that answer Fahnestock's desiderata: the model itself will suggest problems for examination and constrain the development of hypotheses along those lines; the actual rhetorical analysis will be guided and constrained by the limits of the model, revealing weaknesses in the method and suggesting revisions to hypotheses; and the formal structure of the model makes it transferable to other studies by other scholars in rhetoric of science.

Accordingly, my method can be extended to the study of current scientific and technical genres and their readerships. Since it mines reader reactions to produce ranked filters of reader expectations about genre and content, the OT-based model can be combined with traditional ethnographic methods to create portraits of how different readerships—say, engineers and marketers, scientists and humanists, or managers and technicians—evaluate and utilize a particular genre. For example, my method could be used in a project similar to Dorothy Winsor's rhetorical study of the *Challenger* shuttle disaster. Winsor located the "disaster" in miscommunicated expectations among NASA management, the supervisors at the subcontracting firm responsible for testing the problematic o-rings, and the testing engineers and technicians. My method could be used to extract these latent expectations and show how the three groups ranked them differently, thus achieving differing interpretations of memos that clearly stated the failure of the o-rings. Perhaps the managers expected the genre of the memos to match that of a successful progress report, whereas the engineers treated the memos as diagnostics; this could have led to readings at cross-purposes between the two groups, as could differential rankings of the expectations of cost, expedience, and precision when these values came into conflict with each other. The result of adding this facet to a case study such as Winsor's would be a formalization of the researcher's intuitions about divergent professional cultures and precisely how they affect coordinated reading and writing activities.

Reading Science in Nineteenth-Century America

Reader-oriented critics of nineteenth-century American literature currently find themselves in a difficult double bind, according to scholars of historical reading such as James Machor and Jane Tompkins. On the one hand, situated, cognitively focused readings—such as those conducted by Roland Barthes, Wolfgang Iser, and reader-response critics such as Stanley Fish and Jonathan Culler—produce compelling accounts of individual processes of interpretation but still take the text as their "primary unit of meaning," thus sacrificing a cultural perspective that would reveal the text as a "force exerted upon the world."[13] On the other hand, new-historical studies of reading in the nineteenth century yield rich cultural contextualization, but as a practitioner of this methodology, James Machor is concerned that this broadening of focus often sacrifices the ability to account for how individual historical readers actually read.[14]

Empirical studies of reading are one way out of this Heisenbergian sort of inability to keep both process and reading in focus at the same time. Read-aloud or think-aloud protocols, interviews, and surveys, such as those conducted with scientist-readers by Berkenkotter and Huckin, Bazerman, and Charney, help capture moments in the cognitive act of individual interpretation. By studying multiple readers, these studies can average findings across the group in order to discern common reading strategies.

However, these methods cannot help scholars studying nineteenth-century readers and reading practices. The absence in time and space of the original readers and contexts of reading have led reading scholars either to analyze the reactions of a reconstructed "ideal reader" (usually coextensive with the reactions of the scholar) or to elaborate the reading context and culture through archival and historical research.

Hoaxes present a unique opportunity to move beyond this double bind of reader-oriented historical criticism. They are successful rhetorical experiments in identifying and reperforming common reader expectations. In other words, authors of successful media hoaxes managed to guess correctly at least some of the expectations their readers had about science—how it operated in society and how it was reported. The hoaxers exploited these expectations to produce texts that readers bought as the "real thing"; then they revealed their hoaxes publicly to embarrass readers for adhering so blindly to those expectations. In their guessing game, the authors had the immense advantage of living and reading among the people they were trying to fool. However, in many cases, the hoaxers, their readers/victims, and contemporary historians of the hoaxing events have left us a paper trail of reader responses that showcase certain "sticking points" or recurring topics in the readers' debates over the truth-value of the hoaxes or justifications of their decisions to either believe or disbelieve the hoax stories. These recurring topics or topoi index the expectations about science and science news that the hoaxers were exploiting.

Once we have reconstructed these expectations from all the individual responses and then synthesized them into a ranked "filter" of cognitive assumptions that readers' decisions about the verity of science news passed through (unconsciously or consciously), we have a way of explaining why so many readers read the hoaxes as true stories. The filter should enable limited claims about the culture of reading at the time of the hoaxes while simultaneously providing a framework flexible enough to model individual experiences of the hoaxes at different times and places. The next section is dedicated to reviewing the historical justification for an expectation-based model of reading and detailing the methods for recuperating reader expectations from historical documents and modeling their interaction in reader decisions about the truth-value of hoaxes. While this new hermeneutic does not pretend to fully account for cultural context or for individual process, it does provide an extension of rhetorical methods for situating reader-oriented criticism in a specific historical moment. In doing so, it offers a way out of the double bind of context or process that has dogged reader-oriented criticism of nineteenth-century texts.

This study of hoaxing will make one final contribution to our understanding of nineteenth-century rhetorical and literary practice in America, by raising the profile of a neglected American writer of the period: Dan De Quille (William Wright). Dan De Quille was a miner on the Comstock Lode before he became a successful journalist for the *Territorial Enterprise* in Virginia City, where he worked with Mark Twain. While Twain eventually returned to the east, De Quille remained to become a respected mining writer, one of the West's very first technical communicators, who nonetheless mixed in an occasional scientific hoax with his regular scientific and technical journalism. Lawrence Berkove at the University of Illinois and Richard Dwyer and Richard Lingenfelter at the University of Nevada, Reno, have collected De Quille's tall tales and published them along with several scholarly articles on his journalistic and historical writings. However, very few other scholars have worked on De Quille, and no one has yet taken his hoaxing seriously as part of the impact of science on the American West. De Quille surpasses Poe as a career hoaxer, and any consideration of scientific media hoaxing in the nineteenth century must reckon with De Quille or risk underrepresenting a remarkable strategic move in the literary reaction to science. De Quille's science hoaxes and his historical texts, such as *The Big Bonanza*, played a powerful role in the creation of the idea of the West—especially the conception of the West as a natural and scientific waste-and-wonderland; therefore, his cultural influence requires examination.

Reconciling Literary Criticism and Linguistics

Finally and most important, this project invokes (reinvokes) a philological ideology in the study of literary texts. The method I develop here reunites literary and linguistic modes of textual criticism, long assumed to be incommensurate. The mother of this political move was necessity, as I was forced to turn to both linguistic and reader-response methodologies to answer my primary nonliterary, pragmatic questions I had about the hoaxes. Because hoaxing is a social project that expands beyond the physical text to enlist an author's intentions and readers' knowledge about genres and scientific culture, I needed a method that could take as its basic unit of study a whole cycle of rhetorical interaction among an author, texts, a medium, and a reading community, an expansion of focus that David Kaufer and Kathleen Carley argue is crucial to understanding how media affects culture over time.[15] This new methodology I have developed uses reader-oriented and new-historical heuristics along with optimality theory—a constraint-satisfaction framework borrowed from linguistics. I will explain briefly how I developed my method here and provide a quick guide to reading the optimality theory tables that appear in this book. Readers who have only a cursory interest in the method will then be prepared to skip to the chapter(s) about the hoaxer(s) in which they are interested. For those researchers who may wish to try using my method on problems with reconstructing historical acts of reception, chapter 2 contains a more in-depth explication and justification of it and concludes with a tutorial.

In order to study how the scientific media hoaxes had changed the worlds of their readers when they were published, I knew I needed something akin to Grice's maxims, which I had discovered during my semantics and pragmatics courses as part of my M.A. program in linguistics. H. Paul Grice studied implicatures in conversation, which are assumptions (as opposed to hard-and-fast grammatical rules) that help speakers coordinate activities with each other in the real world. Grice formalized his intuitions about these implicatures in four basic maxims:

- Maxim of Quality: Tell the truth.
- Maxim of Quantity: Be as informative as expected.
- Maxim of Relevance: Make your contribution relevant to what has come before.
- Maxim of Manner: Be brief, orderly, and clear.

While in the majority of prosaic communications—emails, asking for directions, business transactions, and so on—these implicatures are scrupulously observed, they can be deliberately broken to create various conversational and literary effects. Flouting is consensual departure; violation is unilateral departure. To take the example of the maxim of quality (telling the truth), a flouting of the maxim produces sarcasm, as in the following exchange:

A: What did you think of the statistics lecture?
B: Riveting.

B does not think statistics is riveting, and she likely accompanies her response with a peculiarly flat intonation to help clue A in to the fact that she is flouting quality, so he should understand her response ironically. Contrast this coordinated activity with lying, which is a *violation* of quality. If B wishes to tell A a lie, she must give no indication that she is not cooperating with his expectations of quality; that failure to share is the essential condition of A's deception.

The Gricean maxims help structure indirect speech acts and speech play like sarcasm and irony. I knew that something like flouting and/or violation (depending on whether readers gradually caught on to the hoax or were duped by it) of communicative conventions was at the heart of the hoaxes' ability to fool their readers. However, the maxims would not answer my questions about the hoaxes because the implicatures are very general rules governing conversation, whereas writing and reading the hoaxes involved knowledge of certain very specific conventions of the science news genre in the 1800s. What I needed was a similar methodology that would allow me to examine hoaxes as communicative acts that constructed meaning not merely through their texts alone but rather as an interaction of the words of the text with both the readers' and the hoaxers' preconceptions and desires concerning the public face of science. I needed this sort of pragmatic angle to my methodology if I wanted to treat the hoaxes as real historical events that had social consequences at the time of their writing and reading.

Enter optimality theory, a constraint-satisfaction framework that models complex decisions made in the face of multiple competing constraints of varying strengths. Optimality theory (OT) is not actually a theory. It is a model for constraint satisfaction processes in general (such as workflow and decision problems, some cognitive processes, and biological processes such as adaptation). Alan Prince and Paul Smolensky brought optimality theory from economics into linguistics in 1993, where it proved useful for handling complex phonological problems previously inexplicable or oversimplified by generative grammar.[16] I encountered it as a phonological tool. However, OT is now being applied to syntax with a more limited degree of success, and a few studies have even applied OT to pragmatics, using Gricean rules for interaction, though these innovations are recent and relatively speculative. Bruce Hall's "Grice, Discourse Representation, and Optimal Intonation" is an example of this new work. How I used OT to model reader decisions about hoaxes is depicted in figures I.1 and I.2. This particular reader was responding to Richard Adams Locke's story series in the New York *Sun* about British astronomer J. F. W. Herschel's supposed discoveries of life on the moon. The series began on August 25, 1835, and ran for about a week, during which time reader responses were printed in the *Sun* and in rival New York papers. This reader noted several scientific and logical inconsistencies in the story but then qualified his objections.

What we see in figure I.1 is the reader's suspicions about the plausibility and internal coherence of the story competing against his desire not to besmudge the reputation of the astronomer; the struggle is indexed by the negative polarity items "however" and "wrong." In the end, this reader's respect for Herschel overcomes his doubts, and he chooses (provisionally) to believe the story.

Modeling a dilemma such as this is straightforward in optimality theory. The decision is presented as a matrix. The top row is for the preconceptions the reader brings to the interpretive process. In this case, these preconceptions are that Herschel is a trustworthy authority, that science

"After all, however, our **doubts and incredulity** *may be a wrong to the* **learned astronomer**, *and the circumstances of this wonderful discovery may be correct. Let us do him justice, and allow him to tell his story in his own way."* = **"true"**

Figure I.1 Graphical representation of reader expectations at stake in a reader's decision to believe the Moon Hoax based on JFW Herschel's scientific authority

	Authority	Plausibility	Internal coherence
✓TRUE		*******	*
FALSE	*!		

Figure I.2 An Optimality Theory Model of the Reader's Decision in Figure I.1

news should be on plausible topics, and that science news stories should be internally coherent. The column down the left lists all likely interpretations of Locke's story. Since the decision we are considering is whether or not to believe a hoax, the interpretive options are simplified to "true" and "false." Figure I.2 as it stands represents the output state of the model, with authority ranked above plausibility and internal coherence (the ranking is signified by the solid black vertical line). To get to this conclusion, which tells us that the reader weighed Herschel's reputation as more telling than his own misgivings about the story, I worked backward from his decision about the story ("true"; the decision is signified by the check mark) and from the ways in which the story satisfied or violated each of his preconceptions about science news. My goal was to end up with a ranking of his preconceptions—a picture of what mattered most to him when he read science news.

Representing the decision involved careful attention to the particular interactions between the story and the readers' preconceptions about science news. Locke's hoax contained, according to several readers' counts, at least eight implausible details and one violation of internal coherence (a change in the claimed optical power of Herschel's telescope). These violations are indicated by nine asterisks (*). To still believe the hoax as "true" would then require an overlooking or outweighing of these nine gaffs. And indeed, that is what we see in figure I.2. The reader's chosen interpretation of "true," indicated by the check mark, reveals that his respect for Herschel's reputation as a scientific authority defeats the evidence of implausibility and incoherence in the moon story, a result that concords with the intuitive assessment of the response depicted in figure I.1. If the reader had chosen to *dis*believe the moon story and cry "Hoax!" his decision could be read across the columns in the "false" row. In this case, his conscience would have been clear in terms of his need to believe only science news stories that were plausible and internally consistent (this one *was not*); however, he would have had to publicly impugn Herschel's reputation, which was an unacceptable violation of his preconception that scientific authorities tell the truth. That unacceptable violation is indicated by the exclamation mark (!). In the end, it knocks the "false" interpretation out of the running in the contest and secures the position of the authoring constraint as stronger than the constraints of plausibility and internal coherence, as visualized by the bold black line dividing them. The dotted line between plausibility and internal coherence indicates that they do not compete with each other but rather work in concert in this reader's decision to believe the hoax.[17]

This is just one of the thirteen interpretive "games" I have constructed in the course of this study of forty-four contemporary reader reactions to eleven scientific media hoaxes published from 1835 through 1880. Synthesized, the results of these games allow me to hypothesize a ranked "filter" of expectations that antebellum and postbellum readers of science news in America brought to their individual reading experiences. Without losing sight of the individual reading experience, I am able to make limited but important hypotheses about science newsreading culture in two periods: 1835–1849 and 1862–1880.

Optimality theory, combined with a new-historical/rhetorical method for reconstructing reading expectations, has helped me answer my questions about hoaxes. I have learned how the hoaxes worked—by performing reader expectations about science news. And I have learned why their authors chose to write them—to surface powerful, unstated assumptions about science's claims to be the new oracle of truth in American public life. Since this is the maiden voyage of optimality theory in literary waters, I dedicate a significant portion of chapter 2 to justifying my choice of the method in the form of a critical genealogy, explicating the application of the method, detailing problems that arose in that application, and answering challenges from critics in literary and rhetorical studies to my "importation" of pragmatic-linguistic methods into this project.

Chapter Summaries

In chapter 1, I lay the historical foundation for my redefinition of hoaxing and set up the kairos or exigence for the outbreak of scientific media hoaxing in antebellum America. After a more thorough explication and defense of my methodology for studying the hoaxes in chapter 2, I then turn to the analysis of individual hoaxes. In chapter 3, I model contemporary reactions to the hoaxes of Poe and his media rival Richard Adams Locke. I develop a preliminary "filter" of ranked expectations held by antebellum science newsreaders. A careful examination of the rhetorical process by which a hoax creates a double readership—dupes and savants—when matched with Poe's cosmology expressed in *Eureka*, reveals that Poe used hoaxes not just to demonstrate his superiority but also to materialize a community of like-minded savants who rejected the "illusion" of professional Baconian science in favor of an epistemology of imagination.

In chapter 4, close examination of reader responses to The Petrified Man (1862) and Twain's other nonscientific hoaxes leads to changes to the filter of reader expectations in order to reflect changes in newsreading culture since 1849. Twain's commentary on his hoaxes offers new insights into the psychology of hoax reading, specifically the power of attention and of reader agendas in constructing belief or doubt. Twain's scientific hoaxing emerges as a special social mechanics engineered to produce laughter as an affirmation of self-determination, engineered, also, to demonstrate Twain's considerable authority over his readers. These findings encourage revision of traditional characterizations of science as an antisocial, mechanistically destructive theme in Twain's work.

Chapter 5 constitutes the first extensive rhetorical assessment of the hoaxes of miner and writer Dan De Quille (William Wright). De Quille's hoaxing is unique in that he embraces both science and the power of popular science writing, including hoaxes, to literally build worlds, to construct the state of Nevada and the idea of the West. Critical of eastern commercial appropriations of western resources, De Quille enthusiastically championed self-made scientists and engineers as the new folk heroes of the West.

In chapter 6 I answer my final question about these hoaxes: What is a hoax? Based on all of the hoaxers' preoccupation with machines, I propose the hoax as a rhetorical machine that transforms public assumptions about science into awareness that scientific truths are constructing a new reality for nineteenth-century Americans. In the conclusion I analyze the Sokal hoax as yet another move in the construction of a tense relationship between American arts and sciences in the media that began with the nineteenth-century scientific media hoaxes.

One important note on the scope of my study: physical hoaxes—faked fossils and artifacts—proliferated in the period under examination here, and along with the media hoaxes, they constitute a text of both cooperation with and resistance to the project of "real" science in antebellum America. Hoaxing is unquestionably a factor in the development of American science, along with the gradual expulsion of the "pseudosciences" such as mesmerism, alchemy, and phrenology from the ranks of the professional sciences. These dynamics are intriguing but too complicated to do justice to in this book. Alexander Boese's trade book, *The Museum of Hoaxes*, addresses to some extent the interaction of hoaxing with the scientific community. James Cook's new book, *The Arts of Deception*, connects hoaxing with antebellum fascination with authenticity and fraud. In order to keep in focus and treat with rigor the complex of relationships that hoaxes construct among authors, media, and readers, I have restricted my focus to scientific media hoaxes, journalistic accounts of scientific discoveries or technologies that seemed factual to many readers but that were later revealed to be authorial inventions.

Chapter One

A Brief Natural History of Hoaxing

The word *hoax* is an industrial-age addition to the English language, according to the second edition of the *Oxford English Dictionary*; it first appeared in 1808, just a decade or so before the scientific hoaxes in question began to appear.[1] But the roots of the word can be traced back about two hundred years earlier to the phrase *hocus pocus*, apocryphally considered a parody of *hoc est corpus*, which a Catholic priest would intone during the Eucharist as the host underwent transubstantiation. In this section I undertake a brief survey of famous rhetorical exchanges that have been recorded historically as hoaxes. By accepting and analyzing this folk classification to see how it demarcates hoaxes from closely related genres, I will arrive at the following list of essential hoax features that must be accounted for by my new rhetorical definition.

- Treatment of particular societal tension(s)
- Resistance to closure
- Parasitism on other genres
- Display of genius of hoaxer
- Construction of agonistic relationship between author and reader
- Argumentation at the stasis of existence
- Effacement of textuality
- Destabilization of reality
- Construction of insider/outsider dynamic
- Division of audience according to differing world views
- Dependence on news media

These features will all emerge during the following historical analysis, beginning with the first recorded media hoax, by Jonathan Swift, which clarifies the differences between hoaxing and satire.

Swift's Hoax and Satires

Alexander Boese's *Museum of Hoaxes* provides the most complete chronology of Anglo-American hoaxes currently available. The first published hoax on his timeline is a fake almanac by Isaac Bickerstaff in 1709. Bickerstaff, better known to us as Jonathan Swift, predicted the death of famous astrologer John Partridge and backed up that hoax with a fake obituary for Partridge printed on the day he was supposed to have died. Swift supposedly concocted his almanac to embarrass Partridge publicly, and indeed, Partridge stopped publishing his own astrological almanacs for a period of six years after the hoax.[2]

Contrasting the Bickerstaff almanac with Swift's later inventions, *Gulliver's Travels* and *A Modest Proposal*, helps distinguish hoaxes from satire. All three works were published widely and anonymously (the first Irish edition of *A Modest Proposal* was signed "Dr. Swift," but the English editions were not). All three were designed to publicly humiliate a person or group of people. But the latter two were satires; they could not have been taken seriously past a few sentences' reading, the one espousing cannibalism, and the other introducing talking horses. The hoax almanac, however, was meant to be believed by readers and was believed, as Partridge himself reportedly learned after a local priest knocked on his door the day of Swift's phony obituary to consult on funeral arrangements.[3] Two groups were meant to be embarrassed by the almanac: Partridge and other astrologers, on the one hand, and the gullible readers who believed in astrology, on the other. The readers, by believing the almanac, became unwitting targets of Swift's two-pronged attack.

This central difference between Swift's satires and his hoax, hinging as it does on the role of the reader, points out that distinguishing a hoax from a satire is almost impossible at the level of the physical text, because a hoax shares many textual characteristics with satire. Dustin Griffin's *Satire: A Critical Reintroduction* redefines satire against its traditional classific-ation as a comedic genre that offers its readers criticism of elite classes and standard mores, catharsis for potentially explosive social tensions, and a satisfying sense of closure. Griffin claims that, in reality, satire is more complicated, deconstructing the "safe" critical distance it offers its readers even as it constructs it.[4] Four textual hallmarks of satire, according to Griffin's poststructural redefinition, apply to hoaxes as well: controversial topics, resistance to closure, parasitism on other genres, and display of genius. I will examine these similarities first before explicating the differences between satire and hoax.

First, hoaxes and satires are both strategies designed to redress power imbalances between conflicting cultural factions:[5] conservatism versus liberalism, elite versus middle class, or in the case of Swift's hoax, science versus astrology. Second, although satires are responses to entrenched cultural programs and values, the satire itself is resistance, a guerilla tactic of exposure and explosion, not a method of achieving closure. Closure is superimposed on the satire by readers with counter-establishment agendas.[6] Thus, a satire such as *A Modest Proposal* is not really a proposal or solution at all. Rather, it performs the cruelty of the establishment (British land-owners in Ireland) without offering any strategies for redressing the grievances of the Irish;[7] those strategies must be brought to the reading experience by Irish reformers and other readers who subscribe to antiestablishment ideologies. Similarly, hoaxes also refuse to tie controversial issues up neatly for their readers. For those readers who "fall for it," the last stroke of a hoax such as Swift's almanac is to embarrass them by revealing itself to be a fake. Once the hoax has thus embarrassed its readers, it is done. It offers no closure, no antidote or resolution to their discomfort. It does not tell them how to stop believing in astrology or what to believe in instead.

Third, Griffin points out that a satire such as *Gulliver's Travels* has a parasitic relationship with the textual genres it imitates,[8] popular travel narratives in this instance.[9] A satire makes fun of a genre or a person by exaggerating the contours of its target's conventions or character. The reader of the original genre recognizes both the correspondences between the target genre and the satire, and the departures; the gaps provoke the laughter, a reaction to lack, desire, difference. This same dynamic certainly holds for a hoax such as Swift's almanac, which targeted and imitated perhaps the most widely read genre of the time.[10]

Finally, satire is designed to display the genius of the satirist.[11] So is a hoax, which is one reason why revelation is so crucial to the hoax's effect on the reader. Nothing in Swift's text revealed it to be a hoax; rumor later outed Swift as the author of both the almanac and the obituary. Undoubtedly, the reputation as a wit that this hoax and his other satires built for Swift must have motivated him powerfully, for his indirect criticisms brought him censure and even imprisonment. However, what is interesting for this project is the fact that a huge part of the action of Swift's hoax—the revelation—occurred outside the text, which is where we must look in order to distinguish hoaxes from satires.

To tease apart the rhetorical effects of these two genres, it will be helpful to apply the approach of Kaufer and Carley and consider not just the texts of satires and hoaxes, but their status as events that instantiate communicative communities—communities comprised of an author, readers, a medium, a topic/issue, and groups indirectly influenced by the communicative event. From this perspective important disjunctions between satire and hoaxing appear. Most important, a hoax is distinguished from a satire by its singling out its readers for criticism—not just Parliament or Irish landholders or an astrologer. Unlike a satire, which constructs author and audience as united in an act of indirect social criticism, a hoax constructs an agonistic relationship between readership and author. The whole point of a hoax, in revealing its artifice, is to embarrass its audience into admitting the inconsistency or poor foundation of its assumptions about what holds true in the world—much like the crux of instructive embarrassment or *elenchus* that was the goal of Socrates's dialectic method.[12] Hoaxes can of course have educative results, but their refusal to offer their embarrassed readers closure by telling them what they can do to alleviate their embarrassment limits further comparison with Socrates's method.

A second distinction between a satire and a hoax is that they are arguments at different stases. Stasis theory is a classical system for structuring forensic (courtroom) arguments, adapted by Jeanne Fahnestock and Marie Secor to the analysis of scientific, civic, and literary discourse. The ancient Roman legal system recognized levels or stases of inquiry into a case that are reminiscent of the "Who, what, when, where, why, how" guidelines of journalistic presentation. "What happened, if anything?" provokes argument at the stasis of existence. "What sort of thing was this happening?" takes the argument to the stasis of definition. "What are the causes of this happening?" addresses the stasis of cause. "Was this a worthy or an unworthy action?" promotes the argument to the stasis of evaluation. And "What should be done about this situation?" brings the argument finally to the stasis of action.[13] A purely text-based, nonrhetorical view of satires and hoaxes might rank them both as evaluative arguments. But only a satire is principally an evaluative argument, designed to call into discussion the goodness or badness of a person, style, genre, or policy; a hoax is an argument at the stasis of existence, playing on the question of whether some happening—or, actually, a reliable witness to that happening—holds true in the world inhabited by the hoax's readers. In other words, what Swift's readers were worried about initially was the question of John Partridge's mortality, not his value as an astrologer.

Certainly, after Swift's reader was embarrassed for falling for the trick, a sort of evaluation could be inferred from that embarrassment: "Believing something just because it claims to be astrology is stupid." But that is an indirect rhetorical move of the hoax; the direct move is always to call reality and its construction into question. By contrast, "satire proper," according to Griffin, "rarely offers itself as 'objective' or documentary . . . Alerted by its generic signals, we are not likely to mistake a satire for fact, not likely to overlook its avowedly 'rhetorical' nature."[14] And indeed, Swift's satire *A Modest Proposal* alerts its readers early on that it is not to be taken seriously:

> I shall now therefore humbly propose my own thoughts, which I hope will not be liable to the least objection.

> I have been assured by a very knowing American of my acquaintance in London, that a young healthy child well nursed is at a year old a most delicious, nourishing, and wholesome food, whether stewed, roasted, baked, or boiled; and I make no doubt that it will equally serve in a fricassee or a ragout.[15]

The awful shock of Swift's implausible cannibalistic proposal steers its reader away from taking it seriously; instead, the reader makes the brunt of her angry revulsion the "cannibalistic" behavior of the Irish landlords. A hoax such as Swift's almanac works very differently. It crucially counts on at least a large percentage of its readership indeed "overlook[ing] its avowedly 'rhetorical' nature" and taking it seriously as the true report of Partridge's demise; if they do not, they do not put stock in astrology and thereby prove immune to Swift's attack later when his almanac is revealed to be bogus. The locus of the effect of a hoax is always in the reader. A reader who believes a hoax such as Swift's almanac, or Locke's reports of moon bison, actually inhabits a different world—constructed by her new beliefs about what is possible in that world—from the world of a reader who "sees through" the hoax and reads it from a skeptic's perspective. Thus, hoaxes build different epistemological worlds for different readers, and the whole raison d'être of the hoax is to embarrass its readership for its misapprehension of the "real" world.

Parody

Eighteenth-century Enlightenment media were also fertile ground for parodies, such as Pope's *Rape of the Lock.* Is a hoax just another form of parody, since, as pointed out above, a hoax must mimic whatever text it purports to be a true example of—whether a travel narrative, almanac, or science report?

I will argue that these genres also differ, this time on grounds of mimesis. A hoax destabilizes reality for readers, calls into question the ways in which they verify that the world they believe in is the "real" one. Therefore, anything in a hoax's style that calls attention to its textuality—such as hyperbole or punning, for example—is at least an initial hindrance to its rhetorical purpose of altering readers' realities. Moreover, attention-getting textuality is the hallmark of parody and burlesque. For these genres to achieve their critical effects, the reader needs to recognize them as texts mimicking other texts, either a whole genre of writing or a particular author's style. The *Rape of the Lock* was only funny to readers already fed up to the gills with bad epic poetry: they were familiar with the various rhetorical features Pope employed to puff up an inconsequential topic (the snipping of a lock of hair), such as the Invocation to the Muse and deus ex machina. Pope's exaggerated mimicry of these features constituted the bite of his poem. A century after Pope, Edgar

Allan Poe's burlesques, such as "How to Write a Blackwood Article," "A Predicament," or "Loss of Breath," similarly focus reader attention on the hallmarks of the gothic "Blackwood" fiction. Consider the opening sentences of Poe's burlesque "A Predicament":

> It was a quiet and still afternoon when I strolled forth in the goodly city of Edina. The confusion and bustle in the streets were terrible. Men were talking. Women were screaming. Children were choking. Pigs were whistling. Carts they rattled. Bulls they bellowed. Cows they lowed. Horses they neighed. Cats they caterwauled. Dogs they danced. Danced! Could it then be possible? Danced! Alas, thought I, my dancing days are over! Thus it is in the mind of genius and imaginative contemplation, especially of a genius doomed to the everlasting, an eternal, and continual, and, as one might say, the—continued—yes, the continued and continuous, bitter, harassing, disturbing, and if I may be allowed the expression, the very disturbing influence of the serene, and god-like, and heavenly, and exalting, and elevated, and purifying effect of what may be rightly termed the most enviable, the most truly enviable—nay! the most benignly beautiful, the most deliciously ethereal, and as it were, the most pretty (if I may use so bold an expression) thing (pardon me, gentle reader!) in the world—but I am always led away by my feelings.[16]

Compare this hyperbolic catalogue of tropes typical of the sensational fiction Poe himself wrote for *Blackwood's Edinburgh Review* to the opening of his self-described media hoax Hans Phaall:

> By late accounts from Rotterdam, that city seems to be in a high state of philosophical excitement. Indeed, phenomena have there occurred of a nature so completely unexpected—so entirely novel—so utterly at variance with preconceived opinions—as to leave no doubt on my mind that long ere this all Europe is in an uproar, all physics in a ferment, all reason and astronomy together by the ears.[17]

Certainly both the burlesque and the hoax open with an excited and exaggerated tone. But the burlesque draws attention to its artifice immediately with its ludicrously repetitive hyperbole. Hans Phaall, even though it is far and away the coyest of Poe's hoaxes, does attempt to salvage its guise as a news story with impersonal third-person narration, science journalism jargon such as "by late accounts" and "phenomena," and an implicit argument that the story is true, as it will soon have "all Europe . . . in an uproar."

It is this argument for the truthfulness of the material presented that marks a primary difference between hoaxes and parodies/burlesques. The focus on textuality and/or style in burlesque and parody serves to shift the reader's attention away from the truth status of the events reported in the story; for example, believing there actually was a drowning baby, a heroic diver, or a tortuous affair is irrelevant to appreciating Poe's "Assignation." The story is parodying the Byronic pose and Byron himself.[18] By contrast, what is at stake in a hoax such as Hans Phaall or Swift's almanac, what is salient to the audience and what they must decide upon, is not primarily who is being pilloried in the story, but whether the events portrayed in the story really happened or not. As a result, hoaxes often have a very low-key or dry style in order not to distract the reader from the content. In his study of Twain's and De Quille's journalism in Nevada, Wilbur S. Shepperson identifies exactly this "stylelessness" as the hallmark of the indirect social criticism seen in their hoaxes; the lack of style performed the profound moral and cultural lackings they observed in the mining boomtowns in which they lived.[19] To sum up, what comparison with parody and burlesque reveals about the hoax is that a hoax resists textual definition by effacing (at least initially) its own textuality and authorship.

Nineteenth-Century Fraud, Tall Tales, and Science Fiction in America

The differences in media hoaxing in the hundred years between Swift's and Defoe's hoaxes and the scientific hoaxes that catalyzed this project are striking. Not only are the eighteenth-century hoaxes few and far between, but they are also published in pamphlet form and reflect the concerns of the English at the time with travel and foreign relations. Hoaxes in nineteenth-century American news media, however, reflect the concerns of a new republic that is finally getting up a good head of steam, literally as well as figuratively; thus, industry and technology, politics, and the scientific wonders being discovered on a daily basis on the new continent all loom large in hoaxes of this era. Antebellum hoaxes, in further contrast to Enlightenment media hoaxes, also had at their disposal well-developed print media, including the important additions of the literary monthly and the penny daily. These advances partially account for the proliferation of hoaxing in the decades before the Civil War, as will be discussed shortly. But before we turn to the cultural kairos that fostered the explosion of antebellum hoaxes, it pays to distinguish hoaxing from a final crop of similar genres that sprang up at this historical moment in response to similarly industrial stimuli: the fraud, the tall tale, and science fiction.

Warwick Wadlington in *The Confidence Game in American Literature* pinpoints the midnineteenth century as the heyday of the con man. Certainly several of the same dynamics that favored hoaxing favored fraud: a population boom that forced Americans to start doing business with strangers, whether they liked it or not; a westward-racing frontier that exposed new jaw-dropping astonishments every day with which law enforcement could scarcely keep up; and competition for resources among immigrant groups and socio-economic classes. Why are the frauds these con men (and women) perpetrated not hoaxes, then, if they are responses to similar tensions, and they both involve the duping of large numbers of people? Steven Mailloux, during his analysis of the trope of conning in *Huckleberry Finn*, explains exactly how he believes a fraud goes beyond a hoax: "[T]he confidence man is not interested in simply performing tricks for the fun of it. He plays his game for a reason, seeking to turn rhetorical exchanges into economic ones, to transform impassioned rhetoric into cold cash. The confidence man thus attempts not only to convince, to affect belief, but also to modify actions for his own benefit."[20] These mercenary concerns of fraud are probably the easiest fracture to identify between hoax and fraud. Hoaxers are after their readers' assumptions; frauds are after their cash. Certainly, hoaxers are interested in a payoff, too, in the subscription rates that come with publicity and notoriety. But hoaxers must reveal their hoaxes to embarrass their readers and launch their social critiques. Frauds avoid revelation and hope that the naïve assumptions that encouraged you to give them money will remain in place so they can dupe you again.

A critic of this distinction between hoax and fraud might legitimately point to the first hoax mentioned in the *OED*. The Great Stock Exchange Hoax of 1814, while not a media hoax, was all about money. A man dressed as a British soldier landed in Dover and traveled to London announcing the defeat of Napoleon. It took a few days for Londoners to get word that, in fact, Napoleon had defeated Blucher, and in the meantime, the news of victory caused a boom in the London stock exchange. As it turned out, the soldier was in the employ of two MPs and a financial adviser, who all profited from the spike in stock prices by dumping their shares. The revelation of the trick was the last thing its perpetrators wanted, so it seems this was a clear-cut case of fraud, rather than hoax, but the fact remains that contemporary commentary labeled it a hoax.

What is to be done with this historical assessment? If I declare these contemporaries inadequate rhetoricians and relabel the Great Stock Exchange hoax a fraud instead, I risk stepping off the folk foundations of this definitional project and rendering it circular—a hoax is defined as I define a hoax. What the historical judgment reminds us of is the fact that money and belief are not always such different commodities. The media hoaxes examined in this project were also about money as their authors made a living selling them to newspapers and magazines. The most famous American hoaxer of all, P. T. Barnum, made piles of money by making people want to see for themselves if the Feejee Mermaid were the "real thing" or not.

The best solution to this historical dilemma is to acknowledge two important differences between the goal of my project and the goals and judgments of the 1814 British media. First, my goal is to define a rhetorical genre, while the purpose of the 1814 reporters was to pass judgment on a public crisis. Beginning with the sense of shock and reality inversion apparent in commentary about the Great Stock Exchange hoax, I am continuing to refine that sense into a model of how a historical hoax works *rhetorically.* That disciplinary evolution may actually be mirroring the ontology of hoaxing and fraud in the nineteenth century, pointing up a second difference between the Great Stock Exchange hoax and the more recent scientific media hoaxes. The two phenomena are substantially separated from each other by time, space, economy, and medium. It is probable that as hoaxing proliferated after the 1830s in American newspapers and as both British and American economies expanded to the point where people were forced to trust their money to strangers in shops and banks, hoaxing and fraud became increasingly distinct from each other as people accumulated experience with both forms of industrial-age deception. After all, these two different labels persist in the language today in order to identify two different social activities. In the end the best litmus test, I believe, for distinguishing hoaxing as a rhetorical genre from fraud is the presence of an indirect message. All of the media hoaxes in this book mounted an indirect criticism of the way the American public was assimilating scientific knowledge. By contrast, the Great Stock Exchange hoax (rhetorical fraud) was not designed to send a message but rather to make a quick fortune for its perpetrators.

The boundaries between hoaxes and the tall tales popular on the mid-nineteenth-century frontier are even trickier to nail down than the boundaries between hoaxes and fraud, if that is possible. Tall tales are the oral forerunners of hoaxes. This inheritance will be examined in greater depth in chapters 4 and 5 on the western hoaxers, but for now we can note that both tall tales and hoaxes play on the existence or witness of a remarkable phenomenon and that audience judgments about the verity of this phenomenon can serve to separate knowledgeable insiders in a community from impressionable outsiders. This dynamic holds when tall tales are told by a conspiratorial group of locals to a tourist in order to demonstrate his/her outsider status, as in chapter 34 of Mark Twain's *Roughing It,* in which frontiersmen fool a "city-slicker" lawyer into arguing a fake property-rights case about a landslide that moved one ranch on top of another.

A crucial distinction between tall tales and hoaxes lies once again outside the physical text in the medium of transmission. Tall tales are an oral genre, whereas hoaxes rely on the relative distance and anonymity of print to fool their readers. Also, fooling people is a relatively uncommon function of the prototypical tall tale. Ormond Seavey in his analysis of Richard Adams Locke's Moon Hoax says that usually "both the deadpan teller of the [tall] tale and his impassive listener [are] conspirators against reality."[21] The "conspiracy" aspect of this description of the tall tale implies it is a joint activity between teller and hearer designed to entertain and distract both of them from daily worries. Whether or not the events of the tall tale actually happened is beside the point in an archetypical tall tale such as the "Pecos Bill" tall tales popular in the later nineteenth century where Bill breaks tornados like bucking broncs. A comparison of one of Mark Twain's tall tales with one of his hoaxes illustrates the differing emphasis on truth-value. The authenticity of the talkative old-timer and the lead-burping frog in "The Celebrated Jumping Frog of Calaveras County" is not what is remarkable about the story; the humor of the situation is. The central claim of Twain's news hoax The Petrified Man, that a human being was found petrified outside Virginia City, is a scientific claim whose truth-value must be assayed. Twain also claimed to have had in mind with The Petrified Man the very "unconspiratorial" aims of humiliating the local medical examiner and shaming his readers, to boot, for their naïve fascination with all things fossilized.[22] This is not a conspiratorial group of insiders putting on an outsider but rather a single journalist multiplying a practical joke through the mechanics of print into a hoax that targets his whole community. These comparisons reveal that while a hoax and a tall tale both call reality and its construction into question, the tall tale is an oral genre emphasizing conspiracy, but a hoax operates at the expense of its readership.

Finally, a hoax is not science fiction. The plausibility of this distinction may seem odd at first glance, since the media hoaxes under consideration take scientific and technological topics at the very moment in the history of American literature when the first science fiction stories were being developed. Edgar Allan Poe, in fact, is still considered a pioneer of science fiction as well as a hoaxer.[23] Science fiction, like the scientific media hoax, attests to the ripple effect in literary communities of the increasing social power of science in antebellum America. The function of science fiction is to dramatize both the best and worst case scenarios of allowing science to dictate social policy. Because of this function, science fiction critic Bruce Franklin claims that the genre helps popularize scientific ideas, that is, inculcate them as moral and social values in lay culture.[24] However, since science fiction by definition does not lay claim to being a true witness to the present or future state of science, it differs significantly from hoaxes, which do initially claim to be reports of the real state of affairs in the world. This difference is nearly invisible in the physical text, as a comparison between the language of Poe's science hoaxes with the language of science fiction stories written by his near contemporary, Fitz-James O'Brien, will reveal in chapter 3. Poe and O'Brien wrote stories on the same topics; however, Poe's were published in news media, and O'Brien's are published in literary magazines, so O'Brien's stories never created a public stir over their truth-value. This powerful effect of different expectations about different types of media will help drive our analysis of a hoax's changing interaction with its readership over time and space.

Kairos

As is observable from the history of hoaxing above, the hoax is a relatively recent rhetorical innovation, dating from the eighteenth century. The hoax, then, is an industrial genre, and this label is more than a temporal indicator. To achieve its effect on readers, American scientific media hoaxing had to wait on certain structures of material and social culture that finally snapped into alignment in the 1830s. Hoaxes could only occur in the kairos, or rhetorical opportunity, created when writers felt the need to interfere in the process of scientific truth becoming public truth in America. Principal among these structural elements that opened up the kairos were these two tensions, both intensified by the American Industrial Revolution: the social tension between the cultures of science and letters played out in the media; and the tension between popular and specialized sectors of the American reading public.

Science and Art

Poe, Richard Adams Locke, and the other media hoaxers at the heart of this book represent the mere crest of a wave of scientific hoaxes inundating nineteenth-century America—such as Maelzel's chess-playing automaton, the Kinderhook Plates (mimicking Joseph Smith's golden scriptures), and P. T. Barnum's myriad artifactual hoaxes, including the Feejee Mermaid. All these hoaxes reflected the intense and very public activity of science and technology in American culture. The Industrial Revolution in Jacksonian America fed (and was fed by) a rapid expansion in both theoretical and applied science, especially in the engineering fields and in the natural sciences of botany and geology. The natural wonders of the American continent, continually paraded before the public eye by expeditions like the United States Exploring Expedition in 1838, provided a seemingly limitless body of data for measurement, cataloging, classification, and publication. In addition, publicly visible and useful technological innovations in the first third of the nineteenth century, such as the railroad, paved streets, and gas lighting, created a clamor for more research and development of labor-saving inventions. The "embarrassment of riches" of natural specimens and data—coupled with incessant nagging from citizens, business, and the government to make scientific research pay off for the public—placed a huge burden on American scientists. At the beginning of the century, scientists were either amateur landowners and clergy who had time to dabble in whatever scientific fields suited their fancy or scientists in the employ of universities such as Harvard or Yale, whose time was divided between teaching and keeping up with their personal researches on the side. The pressure of the data and the public eventually became too much for amateur scientists, so they began in the 1820s to specialize and professionalize in order to organize the workload facing them.

The professionalization of American science also had a political agenda, to mount a patriotic, Jacksonian effort to catch up to the older and better-developed European sciences.[25] Gradually, a professional American culture of "science" coalesced—actually a conglomeration of specialized societies in biology, geology, physics, chemistry, botany, and even phrenology and "magnetism" (mesmerism)—and drew scientific activity out of the view of the lay public. Dabblers and amateurs dropped out, unable to meet the expectations of the new scientific societies. These societies began to publish specialized journals for circulation among their membership. Only a few "general" science journals remained to communicate the real business of science to the lay reader, signal among them Yale scientist Benjamin Silliman's *American Journal of Science*. But these journals, too, often employed jargon and assumed a level of education not universally found in the lay readership.

At the same time this withdrawal was going on in scientific culture, a similar mechanism was at work in the culture of American literature. Increased efficiency of both human and machine labor in America created a publishing boom in the 1820s and 1830s as printing suddenly became faster and cheaper. The Koenig steam press, invented in 1823, probably represents the most significant advance in this department, along with the Fourdrinier process of paper making, developed in 1799, and the cylinder press, which the London *Times* began using to increase its production in 1814. All these innovations had a striking effect on American publishing. In 1825, about one hundred magazines were published nationwide. In the next twenty-five years, that number would increase 600 percent.[26] Book publishing, too, went through a growth spurt, especially toward the middle of the century, according to Frank Luther Mott's account in *A History of American Magazines: 1850–1865*. In the years between 1850 and 1862, the number of books printed in the United States increased by 400 percent.[27]

This development of the print industry, especially the magazine boom, was the first major surge in truly "American" texts as compared to the previous American reprints of European texts. Universities and magazine publishers in particular began to see a need for a critical community and apparatus to cull a "quality" American literature from the landslide of new texts. Accordingly, a series of university-funded literary magazines such as the *Putnam Monthly* and the *Atlantic Monthly* were begun and immediately created a readership that was unabashedly Brahministic.[28] Edgar Allan Poe was actually close to the vanguard in this tradition. He abhorred "puffery," the jingoistic tendency he noticed among literary "critics" in the 1830s to claim that anything written by an American author was good simply by virtue of its provenance.[29] As editor of journals such as the *Broadway Journal* and the *Southern Literary Messenger*, Poe became famous for "broad-axe" criticism—reviews that mercilessly cataloged the flaws of American books and called for standards of criticism that would distinguish a genre of American letters from the "rabble."[30] In this way, the publishing industry in America, the writers it paid (off and on and poorly), and the magazine editors who relied on this industry for content to fill their pages began to form their own community just as specialized and perhaps even more openly antipopulist than the professional scientific communities.

Then the trouble started. As a narrative convenience, we may date it from the publication of Charles Lyell's *Principles of Geology* in 1830–1833 and his 1841 lecture tour in the United States.[31] Lyell's *Principles* suggested a new chronology for geologic history, argued against catastrophic events such as Noah's flood as major geologic processes, and argued for Hutton's view that the earth was much older than traditional estimates keyed to biblical genealogies. Lyell created an uproar, not just between clergy and scientists but between and within scientific and literary communities as well; for, to characterize the *Principles* controversy as a mere matter of science versus religion is to overlook the foundation of American public thought in the textual authorities of the Bible and the Word of preachers, writers, politicians, and philosophers. Lyell essentially suggested that Truth was not to be sought in the Word, but in the World, through the seemingly antitextual activities of observation and calculation.

Men and women of letters reacted strongly but variously to this basic claim. Some, including notably Melville and Hawthorne, saw little less than the death of the human soul in scientists' methods. Others, such as Emerson, transformed an initial resistance to scientific methodology into a nearly rapturous embrace—catalyzed by a life-altering afternoon in the natural history collections in the Jardin des Plantes in Paris—of science as a truth-seeking epistemology on par with the Word and the imagination.[32] Scientists, for their part, perhaps sensing an opportunity in the fracas to expand their political power and garner more funding for their research, borrowed the trope of "progress" from a rapidly industrializing society they had helped create. They used it to argue that the way they saw things was simply the way things were headed, and soon Americans would be forced to see them that same way. There was no escaping either Nature or Progress.

In the debate, each side had help. Science had spectacle in its corner. Past the mind's eye of the public danced visions of Louis Agassiz's gorgeous books full of color plates of turtle specimens, P. T. Barnum's natural wonders in his American Museum, and public exhibitions such as the microbes visible under a new "hydro-oxygen" microscope on a tour of New York museums during 1835. In addition to these tantalizing material displays, scientists could also lay claim to a myriad of technological innovations that their researches authorized, if not actually created. However, these innovations partook in a fierce industrialization of both city and countryside that left many Americans overworked, worn out, and nervous about what machines might do to them. Public literary representatives such as Melville and Hawthorne had this fear on their side when they went public with their criticisms of scientific methods and motives. The legacy of the British romantics, who had mounted their own rebellion against an industrialization that started nearly a hundred years before the American Industrial Revolution, remained strong in the pages of novels, daily newspapers, and sermons delivered from transcendentalist pulpits in the northeastern states. The "machine in the garden," as Leo Marx has termed the presence of technology in antebellum America, was a terrifying as well as a fascinating phantom.

These tensions might be the birth pangs in America of what C. P. Snow deemed the "two cultures controversy" almost a hundred years later, in 1959—a communicative disconnect between the arts and sciences that Snow saw, in the wake of World War II, as a threat to American humanism and democracy.[33] David Kaufer and Kathleen Carley argue that the boundaries between professional communities ossify if they specialize and remove themselves from public oversight, thus exacerbating the problem of interdisciplinary rivalry. Increasing the permeability of boundaries, like the interchange Snow advocated among his literary and scientific friends, reduces confrontation over differences of values and epistemology between professions.[34] Nothing of this sort of rapprochement transpired in the battle following the publication of Lyell's *Principles* in the 1830s. Instead, public literary intellectuals used scientific media hoaxes to mount an attack both against scientists and against the publics who (perhaps unwittingly) supported scientists' campaign to ground America's social policy in scientific values. As we will see in the chapters on the individual hoaxers, the hoaxes were a wrench in the gears of the popularization of ideas such as Lyell's. Exploiting the public's neophytic faith in the truth and beauty of science, the hoaxes—through their dual mechanism of deception and revelation—were able to transform those assumptions into a humiliating self-awareness. The hoaxes coerced readers into admitting the foolishness of their tendency to believe anything that came stamped with the imprimatur of "science." Indirectly, the hoaxes also critiqued the scientists whose work they mimicked; in many cases the hoaxes implied via their counterfeit that the scientists' publicizing of their work—if not the work itself—was also counterfeit.

Popular and Specialized Reading Culture

This critique by public literary figures of the mounting social power of science would not have been as effective if the hoaxers were not also able to exploit their readers' appetites for and trust in the popular media. The withdrawal of both scientific and literary discussion into specialized journals and professional societies left the lay public hungry for news of what was going on behind these closed doors and covers.[35] A uniquely Jacksonian social dynamic of distrust intensified this desire for knowledge and control—a fear of elite, undemocratic repositories of power hidden behind the rapidly bloating federal government, a fear that manifested itself in the 1830s in the persecution of the Masons and the disbanding of the Second National Bank.[36]

Into this tense rhetorical vacuum stepped the genre of the popular science article, pacifying the public appetite for the most sensational of the current scientific discoveries and technological inventions with bold headlines and lots of engravings. The penny dailies sported many specimens of this new genre, and publications dedicated solely to the edification of the popular or general science reader sprang up, including the *American Journal of Science* (1818) and later the *Scientific American* (1845). These journals and papers printed renowned naturalist Louis Agassiz's latest discoveries about glaciation on their front pages but were equally likely to showcase interviews with famous phrenologists and mesmerists and accounts of hay bales mysteriously levitating into the clouds.[37] Catering to an audience hungry for scientific wonders and technological labor- and health-saving gadgets, these ready media platforms created the perfect stage for the scientific hoaxing of Poe, Locke, Twain, and De Quille.

Public desires constitute a powerful force driving both the form and the function of communication between scientists and lay reading communities. In *Counter-Statement* Kenneth Burke claims that any given rhetorical form both creates and satisfies desire within the reader, a desire—in the case of the "gee whiz" popular science articles of the 1830s—for identification with or control over the often alien social force of science and technology. Steven Katz adopts Burke's definition of rhetorical desire to argue that this desire for identification with science has led in this century to scientific discoveries being portrayed as epic quests and scientists as heroes.[38] Dorothy Nelkin in *Selling Science* finds this dynamic operating even as early as the 1890s, as popular science articles portrayed science as a "mystical" knowledge open only to nearly superhuman scientist initiates.[39] Extrapolating this trend back a few decades to the 1830s as the public watched science retreating behind closed society doors, we could argue that the brand new popular science genre was simply reinforcing a Burkean loop of desire already present in the reading culture. The public desired canals and railroads and medicines to make their lives easier, and this desire drove scientists in the form of a constant social pressure; however, scientists' discoveries and inventions also sparked desires within the public for "better, faster, more" of everything.

The popular science article also represented an important transference of trust to the popular media, a shift that paved the way for the hoaxes. Newspapers proliferated in the Jacksonian era as the population in the States expanded to the point where it was impossible to witness directly what was happening in one's own community, much less in Virginia or New Hampshire. Readers came to rely on the news and the mail as vicarious witnesses to important social or political happenings. The political reporting during this time, in particular, reflects editors' awareness that they were performing an experience of virtual witness for their readers; verbatim reports of proceedings of Congress take up pages and pages of newspapers and party-published monthlies such as *The American Review: A Whig Journal*. If readers wanted the information they needed to vote appropriately and to make decisions that affected their families, they had to sacrifice eyewitness and personal credibility and to put their trust instead in the institution of the newspaper and the forms of its anonymous articles. Miles Orvell argues that this coercion of trust was reinforced by a mechanical model of social economy becoming increasingly current in America with industrialization. In *The Real Thing*, Orvell details the fascination of Jacksonian Americans with facsimiles produced by machines and argues that facsimile became an increasingly powerful trope for understanding social and commercial relationships. Stereotyping became a common way to deal with unknown social groups, as Americans adapted the model of machine replication to their social relationships. They became more and more apt to judge what they had not experienced as a carbon copy of their previous experience.[40]

A further consequence of this copying mechanism in the rapidly expanding social economy of antebellum America was that transactions with institutions were gradually substituted for transactions with acquaintances—such as familiar local shopkeepers or bankers.[41] Trust in people had to be shifted to trust in corporations and rules of operation. This shift, forced as it was by the material conditions of a rapidly expanding urbanized environment, created a deep unease in the public consciousness. This unease was performed in the protests mentioned above against the Masons, Rosicrucians, and National Bank; however, an industrialized corporate economy was a fait accompli. Even if they wished to, Americans could not shrink their society down, take the machines out of it, put things back to the way they were.

The scientific media hoaxers took shameless advantage of this coerced trust. They identified and replicated in journalistic form their readers' desires for science, technology, and mechanical facsimile. By giving readers what they wanted and then pulling the rug out from under them, so to speak, the hoaxers confirmed their readers' fears that they were being duped. In fact, the defining feature of a hoax is the moment of embarrassment.[42] In this moment the hoax reveals its devices, which amount to the reader's own assumptions, which the hoax has exploited to achieve its humiliating effect. This revelation can come either within the reading experience or in its immediate context: Twain's hoax The Petrified Man revealed itself textually through sly details revealing that the petrified corpse was thumbing his nose at the reader; Poe revealed his Balloon-Hoax of 1844 within the reading context, by getting drunk and standing on the steps of the Sun trumpeting his forgery to potential subscribers. In either event the revelation crucially depended on the reader's trust in the newspapers' vicarious witness of the "real world." American society had gotten too complex for readers to be able to verify for themselves everything they needed to know in order to function in it. The hoaxes thus constitute both a sharp criticism by literary intellectuals of this state of affairs and a voicing of a deep public uneasiness with it.

Chapter Two

Method

As a reminder, my first two questions about the nineteenth-century scientific media hoaxes were these: How did the hoaxes fool their readers? Why did Poe and Twain and the other hoaxers write them? These questions were essentially pragmatic, and most of the work on the hoaxes that had been done previously was literary. So I went searching for some critical interface between pragmatic-linguistic and literary methods that I could use as a platform for talking about how hoaxes worked culturally and on individual readers. The search forced me to confront a well-known traditional disciplinary genealogy that argues for the isolation of linguistic interests from literary ones based on political history; however, the search also led me to the discovery of an alternate or subversive genealogy that tells a story, against the traditional story of schism, about how certain linguists, critics, and rhetoricians have maintained the basic tenets of an essentially philological program over almost three hundred years.

The following is a brief sketch of the traditional story about how linguistics and literature came to rest in separate departments with separate scholarly programs. Fuller accounts can be found in R. H. Robins and Julie Tetel Andresen's history of linguistics in America and in Gerald Graff's institutional history of English departments in this country. Jacqueline Henkel's *Language of Criticism: Linguistic Models and Literary Theory* exactly addresses the interaction between the disciplines, and my account relies perhaps most heavily on the good research she has done.

The Traditional Genealogy

Both the traditional and the subversive genealogies of linguistic approaches to literature begin with philology. The word *philology* was first used by Plato, most likely in the *Laches*, to denote a love(r) of debate or discussion. The term was revived in 1777 by Friedrich Wolf, a German scholar who used it to define the field he wished to study, overall an "'attention to the grammar, criticism, geography, political history, customs, mythology, literature, art, and ideas of people."[1] This ambitious program would eventually engender both American linguistics and American English departments.

American universities in the nineteenth century were founded mainly on the German model, so German architectures of language and literature studies were imported to institutions such as Johns Hopkins, the home of the *American Journal of Philology* and the *Modern Language Notes*, the precursor of the *PMLA* and the flagship publication of the MLA, founded in 1883.[2] Literary studies were at first language studies, and texts were primarily treated as data to support *Junggrammatiker* or neogrammarian theories of historical sound change in English.[3]

Linguistics and literature began to separate into two disciplines in the 1870s, but in order to understand the reasons for this schism, it is important to appreciate the forces that kept the two disciplines in the same department at most schools until the 1890s. The most important of these motivations was scientific justification. As Gerald Graff puts the case, "The philologists had solved the problem that had perennially thwarted the claim of English literature to be a classroom subject: that you could not examine in it."[4] Philology in the hands of the neogrammarians was something that Plato and Friedrich Wolf would not have recognized. It was mechanical, dealing strictly with what could be empirically demonstrated. It no longer made time or space for musings about culture and mythology. It was rule-driven, testable, and quantifiable.[5] It therefore looked like science, and in an era when land-grant universities were standing or falling largely on the merits of the technologies they could contribute to the juggernaut of American Imperialist capitalism, English departments eagerly hoisted philology up their flagpoles.

However, a competing school coalesced within these departments, one equally motivated by capitalist concerns. These "generalists" realized that the majority of the students they trained were going on to be businessmen in an increasingly global economy, not philologists in a university. These students needed much more than a command of English grammar; they needed a command of Anglo American culture and arts, as these were just as much in currency as the dollar in developing colonial countries eager to stay on the winning side of American global expansion.[6]

These were the political seeds planted in English departments when the first and second World Wars came through and redrew disciplinary boundaries. Philologists dissociated themselves from the German model after the horrible results of the Aryan cultural reconstruction in Europe and, to avoid association with any future disastrous experiments in nation-building, plunged themselves into the minute mechanical description of Native American languages. The LSA was founded in 1924, and linguistics departments began to be formed. Literary studies too abandoned the old comfortable arguments about the inherent global superiority of European languages and literatures and cloistered itself in the New Criticism, in the organic unity of the text. By this point, the job of teaching "proper" American English had fallen through the crack between the splitting disciplines and ended up largely in the laps of high school teachers and composition scholars in the Current Traditional school.

From the 1940s onward, American linguistics had no more use for literature, with few exceptions. It was bent as a discipline (still is) on garnering the distinction of a science, and the only way to accomplish that was to abjure literary aspirations. As Jacqueline Henkel explains, "In the United States . . . opposition between science and literature [was] firm and long standing. Scientific methods did not mesh with New-Critical perspectives; linguistic analysis was at odds with texts conceived as organic and literary language cast as emotive."[7] The story from that point up to today is rather one of literary critics periodically borrowing from and then abandoning linguistic methodologies. The structuralists obviously had linguistic interests, many inherited from the Prague school, whom I will discuss in a moment, but overall they did not engage with contemporary linguistic theory; what they shared with linguists—a focus on language—was so broad as to be useless as a functional bridge between the disciplines. The most protracted instance of disciplinary interaction was the "importation" of Chomskyan generative linguistics and speech act theories by reader response critics in the 1970s and 1980s.[8] Most attractive to these critics—a group including Stanley Fish, Jonathan Culler, Jane Tompkins, and Wolfgang Iser—was Chomsky's essentially cognitive methodology, his description of linguistic competence as a shared, innate faculty, and his positing of an "ideal speaker" against whose competence the well-formedness of utterances could be tested. Some of Chomsky's more technical proposals, however, such as transformational/generative phrase structure grammar, did not make the crossing to literary shores quite as smoothly.[9]

Jacqueline Henkel marks the final blow in terms of serious, programmatic collaborations between linguists and literary theorists as the advent of deconstruction.[10] Derrida singled out three important linguistic progenitors for castigation: Ferdinand de Saussure, the reputed father of the notion of language as a communal rather than an individual possession; and J. L. Austin and John Searle, who developed speech act theory as part of a subfield of linguistics called "pragmatics," which studies language-in-use. Derrida's criticisms of de Saussure in *Of Grammatology* are the most significant in characterizing the break between linguistics and literary theory. Derrida claimed that de Saussure's decision to privilege speech over writing as the "primary" modality of human communication was arbitrary because speech contained writing, and vice versa. He also declared linguistic-scientific methods unsuitable to the study of literature or culture because science could not examine its own foundations; any methodology for studying art and culture had to be able to question its own position in order to correctly read cultural signs. Henkel points out that in spite of this round-sounding rejection, Derrida in fact was not consistent in what he meant by "science," and he made it clear elsewhere that he expected scientific and empirical projects to continue in our society; in fact, he wished them to.[11] However, as deconstruction and its attending postmodern doctrines took hold of literary studies across the country, the auspicious moment for collaboration between linguists and critics passed with some finality.

That is the traditional genealogy of linguistics and literature in America, and in my experience, it holds true on a broad disciplinary level. I found it hard to study both linguistics and literature in either department. The very few linguists who had once published analyses of literary texts hurried to add, "I don't really do that anymore," as if that work were a tattoo or some other slightly regrettable peccadillo from younger years. The literature folks glazed over at the merest mention of words like *phrase structure* or, worse yet, *grammar*. Andresen distills the dynamic that accounts for this schism down to three foundational assumptions by modern linguists: "1) that speech and writing are separate phenomena; for the linguist, speech is primary, writing is secondary; 2) that the linguist, being a scientist, is prohibited from intervening in the mode of existence of his/her object of study; and 3) that language, being a social fact . . . whose subject is the speaking masses, cannot be the object of conscious action, that is, of deliberate change."[12] These proscriptions, inherited from the nineteenth-century neogrammarian tradition with a healthy inoculation of American capitalist technoscience, seem to foreclose on any future collaboration between linguists and literary critics interested in literature as social force. Henkel characterizes the divide similarly from the literary side but in a more dramatic mode worth quoting in its entirety:

> We might pause to consider our disciplinary perspectives on each other, here through the controversy over speech versus writing. Literary critics imagine linguists to be fully committed to the proposition that a logical calculus suffices to explain the complex everyday workings of speech. Linguists (often in introductory texts) invent English teachers who are politically naive enforcers of the standard language in particular and middle-class social norms in general, loyal adherents to unexamined notions of linguistic propriety and correctness. Critics sometimes hint that sociolinguists impose rigid categories on the populations they study. And sociolinguists deplore the presumption that one may write on social issues without conducting field-

> work, without consulting the people one discusses. Critics have objected to linguistics for its simplistic conception of self, for privileging selfhood over textuality. Linguists take persons in English departments to be valorizing the texts of the standard language and high culture over speakers. I think it is safe to conclude that critics and linguists simply do not understand each other's projects.[13]

So much for the traditional genealogy of linguistics and literature. Fortunately for me and my endeavors, it was not the only lineage to be had. My insistence on tracing the pragmatic strain backward up the genealogy of philology's descendents gradually uncovered a second, subversive genealogy of scholars who have maintained a persistent interest in the social productivity of texts.

The Subversive Genealogy

This second genealogy begins again with Wolf's definition of philology as "attention to the grammar, criticism, geography, political history, customs, mythology, literature, art, and ideas of people"—in other words, how words make the world in which we live.[14] The most direct heir of this school of language study in America was William Dwight Whitney. Whitney, the first president of the American Philological Society at its foundation in 1869, taught Sanskrit and comparative philology at Yale. He departed from the dominant German mechanist model in his desire to view language as an essentially political. In Whitney's conception a language lived and worked by the tacit agreement of all of its speakers. Language was not the possession of an individual but a democratic institution that only suffered changes enacted across the board rather than by a few powerful individuals.[15]

It is significant to note that Whitney's emphasis on language as a communal construction predates Ferdinand de Saussure's theory of *la langue* by several decades. In fact, the two scholars communicated with each other, but while de Saussure admired Whitney's ideas, according to Andresen, Whitney appears to have been only an indirect influence on the Swiss structuralists, who were more influenced by French philology.[16]

In spite of Whitney's break from the popular German mechanist model of language study, Andresen argues he was the most influential figure in nineteenth-century American linguistics.[17] That he does not get even a side note in most histories of American linguistics speaks to the power of the traditional genealogy I told you first, a story about mechanical linguistics versus literature in which Whitney must be suppressed. For Whitney, in fact, championed the study of literature. In his view literature was an important conservative force as a language developed and changed, a sort of gold standard for its best and highest uses that was available for public inspection by all speakers.[18]

Whitney's philological heir in America, Roman Jakobson, also unfortunately inherited some of Whitney's inability to keep American linguistics and literature together in any disciplinary sense. Schooled in Russian formalism to pay minute detail to linguistic features in literature, and a founding member of the Prague school, Jakobson campaigned against a view of literature as special language unbound by the constraints of ordinary language use. His legacy is reperformed in studies by structuralists and poststructuralists such as Mikhail Bakhtin, Tzvetan Todorov, Roland Barthes, and Jacques Derrida. However, the Prague school did not make as much headway in America—where the mechanical school determined the departmental politics of linguistics—as it did in Europe, where scholars were still comfortable with linguistic programs that took into account cultural and literary expressions of language.[19]

A contemporary of Jakobson, J. R. Firth, was making similar waves in British linguistics, which had also adopted the mechanical German model. Firth studied phonology, the system of sounds that make up a language, but unlike most of his colleagues, he argued that the first object of phonology was in fact not sound but meaning. Firth deemed inadequate any linguistic theory that was not based on the study of how meaning is made between people.[20] Firth's disciple M. A. K. Halliday has since developed these principles into a rich system of textual analysis that keeps in view the activities that people are trying to accomplish when they speak and write to each other.

Both Jakobson and Firth have had powerful indirect influences on American linguists and critics.[21] Their theories have inspired studies in pragmatics and speech acts by scholars such as Searle and Austin, whose work in turn has been used by reader-response critics, as mentioned previously. Geoffrey Sampson argues that the Prague school's impact was powerfully felt in the formation of modern sociolinguistics in America, typified by William Labov's groundbreaking studies of dialect as a sign that irresistibly invokes class politics.[22] Elizabeth Traugott and Mary Louise Pratt's work on the signification of dialects in American discourse is an extension of this primary concern with the politics of utterance.[23] Finally and important for my project, Firth's meaning-based model of language has influenced not just speech act theory but also relevance theory, a model of communication that I explored as a potential method for explicating the hoaxes.[24]

Toward the end of following this second genealogy of philology, I began to encounter methods that treated literary acts of reading and writing as communicative acts that negotiated meaning as a function of preconceptions held by readers and writers. This was exactly what I was looking for to help me understand how the hoaxes had duped their readers and what the authors were trying to communicate through this process of fooling and revealing. However, as I followed trails of footnotes from one relevant source to another, I found myself traveling in an ever-shrinking spiral among sociolinguistic, cognitive, pragmatic, new-historicist, and reader-response methodologies. The center I was seemingly being drawn toward was rhetoric.

I have just described the sociolinguistic and pragmatic arcs of my methodological search. The reader-response sections of the spiral also follow from previous discussion—the adaptation of speech act and generative linguistic methodologies to the reconstruction of reading acts by critics such as Culler, Fish, and Robert Scholes. Scholes, in fact, put in a few words the very phenomenon I was out to capture in my methodology: "The supposed skill of reading is actually based upon a knowledge of the codes that were operative in the composition of any given text and the historical situation in which it was composed."[25] In order to explain how the hoaxes worked, I needed to get at both of those things: the codes of reading science news that an individual newsreader at that time had in his/her head and the historical context of that act of reading. The problem was that while I found statements like these by reader-response critics helpful and suggestive, many of their actual methods revolved on a key factor unavailable to me—the ideal reader. The readers of my hoaxes, ideal or otherwise, had all died nearly a hundred years before I started studying them. I turned to new-historicist methods in an effort to recuperate my vanished readers, and indeed they helped me identify many of the conditions the readers must have faced as they read the hoaxes in newspapers—down to the quality of the paper the hoaxes were printed on and the light levels they were reading under. But these methods for all their contextual richness could not model the decision of a single reader to either believe or disbelieve Locke's Moon Hoax in the New York *Sun*, which was all I had to work with as data—the individual reactions to this hoax and others preserved in archival sources. In short, after traveling through reader-response and new-historicist territories on my search, then, I found myself in an essentially Heisenbergian dilemma. If I fixed the experience of an individual reader, I lost the context of that reading, and if I fixed the context of the reading, the individual interpretive act was lost. I needed a method that could connect individual communicative acts to their historical and political contexts.

Fortunately, it is exactly this sort of analysis that rhetoricians have gotten very good at over the last couple thousand years. By turning to rhetorical studies of historical reading acts, I found methods that combined the best of reader-response and new-historicist techniques to produce local histories of specific acts of reading. My technique for reconstructing the codes that governing antebellum science news reading was borrowed from three main sources: Charles Bazerman's *Languages of Edison's Light*, Rosa Eberly's *Citizen Critics*, and Steven Mailloux's *Rhetorical Power*. This method writes local histories of specific acts of reading using contemporary reader responses. I applied this method to my own project first by reading the twenty-two extant reader responses to Locke's Moon Hoax and gleaning from them common "sticking points" or topoi that the writers returned to again and again as they argued for the validity of their decisions to either believe or disbelieve the hoax. These topoi included the logical consistency of the story, the novelty of the reported discoveries on the moon, and the reputation of J. F. W. Herschel, the famous British astronomer credited with making them. While each reader made up his or her own mind, all of them made it up with reference to similar topoi. However, that set of readerly expectations was not enough in and of itself to describe how each individual hoax reader had come to his or her conclusion about the hoax. I needed some way of showing how these expectations interacted and perhaps competed with each other to produce belief or doubt.

Two models based on Gricean maxims helped me think about how to solve this problem. These were Dan Sperber and Deirdre Wilson's relevance theory, and Ellen Schauber and Ellen Spolsky's preference rules. Both modeled interpretation as a sort of constraint-satisfaction game in which different expectations come into play and conflict with each other; the resolution of these conflicts produces an interpretation. The problems with the models were that relevance theory was not a reading theory, so did it not allow for effects on relevance arising from the readers' selection of certain textual elements to attend to over others; it also did not provide for the multiple expectations I had already discovered in my research. The problem with the preference rules was that, while they were ranked in terms of importance and separated into linguistic, pragmatic, and literary conventions, the categories could not interact with each other in Schauber and Spolsky's model, and I knew the hoax-readers' understanding of the real world interacted freely with their understanding of the conventions of science news in their decisions about the hoaxes. I needed a way to model the interaction of multiple expectations in an individual reader's decision about the truth or falsity of the hoaxes.

Optimality Theory

In the introduction I gave a brief history of optimality theory and an example of how it can be used to model the interaction of reader expectations in the interpretive experience. Here, I will begin by giving a more in-depth explication of how it has been used in linguistics before showing how it produced unique critical insights in an early application to the problem of rereading in a Jorge Luis Borges short story. In that analysis I demonstrated how specific competitions between expectations—about the reliability of the narrator, for example—can drive a single reader's rereadings of a text. I will give the results of that experiment as examples of my adaptations of optimality theory and then move on to problems I encountered in the application, how they were overcome, and how I answer challenges from literary critics about importing a linguistic methodology into the study of hoaxing. I will finish by directing the interested reader to a brief online tutorial that will illustrate from start to finish how to reconstruct common reading expectations for a particular genre and historical audience.

To recap, optimality theory (OT) is not actually a theory. It is a model for constraint satisfaction processes in general (like workflow and decision problems, some cognitive processes, and biological processes such as adaptation). Prince and Smolensky brought optimality theory from economics into linguistics in 1993, where it proved useful for handling complex phonological problems previously inexplicable or oversimplified by generative grammar. To see how it works in phonology, consider table 2.1.

The first column of the table (called a "tableau" in Prince and Smolensky) technically contains all the possible candidates for the phonological form of a word as speakers actually pronounce it; usually, however, the column only contains logical alternatives—in this case, the three most probable syllabifications of the English word *onset.* A hyphen indicates the syllable break in the word. The bracketed <a> in the third candidate represents a deleted vowel (which is actually a fairly common phonological feature in colloquial English: think of how many times you've gotten a nasal "N-n" as a response instead of "No, no").

The top row of the table lists all phonological constraints applicable to the problem in order, left to right, from strongest (inviolable) to weakest (often violated in practice). In this case, the constraint FAITH, which states that all parts of a word should be pronounced, is ranked higher than both ONS, which says syllables should start with consonants, and NOCODA, which says syllables should not end in consonants. ONS and NOCODA are not ranked with respect to each other because they never operate on the same part of the syllable and therefore never compete with each other; the vertical dotted line signifies this lack of competition. The

Table 2.1 Syllabification of "onset" /ansɛt/

	FAITH	ONS	NOCODA
✓an-sɛt		*	**
ans-ɛt		*	*!**
<a>nsɛt	*!		*

ranking of theconstraints in this table could also be notated in a linear form as FAITH >> {ONS, NOCODA}, where ">>" signifies domination and where bracketing with commas signifies equality of rank and therefore lack of competition. This ranking applies to all English words and was determined via analysis of copious sets of English syllabification data by phonologists.

The asterisks in the matrix of the table represent violations of particular constraints. The violations add up like penalty points against a candidate, with a violation of a stronger (leftward) constraint counting more than a violation of a weaker one. An "!" follows and indicates the fatal violation, the one that knocks the candidate out of the running for optimal form (violations are tallied up from right to left, weakest to strongest). The check mark in the candidate column indicates the optimal phonological form, *the one that satisfies the greatest number of the highest-ranked constraints.* This optimal form is the actual, real-world pronunciation of 'onset.'

In the example in table 2.1, "an-sεt" is the optimal form. While it has more total violations than "<a>nsεt," it nevertheless satisfies FAITH, the highest ranked constraint. The runner-up, "<a>nsεt," does not. The third form, "ans-εt," gets knocked out of the running even earlier because it accrues more NOCODA violations than either of the other two forms due to a consonant cluster "ns" at the end of the first syllable.

How are phonological constraints derived in optimality theory? The constraints and their ranking were derived from looking at phonological paradigms. Paradigms are microcosms of a language: they are data sets listing all unique forms currently in use in the language for plural endings or syllable-breaking patterns or whatever phenomenon is under investigation. Looking at these paradigms, phonologists make general claims based on the patterns they see: "This language never has closed syllables"; "in about half the forms here, consonant reduction occurs when the plural ending is added," and so on. These generalities become constraints on the actual spoken forms of words in the language because the forms that speakers actually use are the ones that break the fewest of these "rules." The total set of phonological constraints, claim phonologists, is universal; only their ranking changes from language to language. In Minyanka, a Niger Congo language, there are never consonant clusters; this is an extremely high-ranked constraint. In English, that is a very low-ranked constraint, as evidenced by five of the words in this sentence. Once phonologists determine which of these universal constraints are important for syllabification, plurals, and so on, in a particular language, their ranking is determined by working backward from the paradigms again. Examining the forms speakers actually use, the "optimal" forms, allows you to deduce which of your constraints are the most dominant, which are middle ranked (coming into play only to decide between two forms that both satisfy a more dominant constraint), and which are very weak (only "winning out" and appearing in actual speech in the absence of any competition with stronger constraints).

Optimality theory (OT) has worked spectacularly well in phonology, perhaps because phonological rules are a relatively circumscribed set, as there is a finite range of perturbations the human vocal tract can perform in the process of combining speech sounds. OT is now being applied to syntax with a more limited degree of success, as the universality of syntactic rules is still actively debated. For one compilation of views, see Barbosa and colleagues, *Is The Best Good Enough? Optimality and Competition in Syntax*. A few linguists have even applied OT to pragmatics, using Gricean rules for interaction, though these innovations are recent and relatively speculative. A recent collection by Blutner and Zeevat demonstrates that a small group of European linguists (and a few Americans) are using OT to solve pragmatic problems ranging from language learning to ambiguous theme marking on Tagalog arguments to speakers' choice of demonstrative forms in Swedish.

Using optimality theory to help organize and model the interaction of reader expectations is a productive addition to expectation-based models of the reading process. In this application the model's constraints are actually very similar to Schauber and Spolsky's preference rules; OT simply adds the benefits of a graphic model, which is easier to inspect visually and which allows both interaction among many different kinds of expectations and a greater degree of precision in the ranking and reranking of those expectations.

OT and Reading an Unreliable Narrator

As an example of how OT can help explain reading expectations at the level of genre, specifically, expectations about the reliability of the narrator of a short story, consider Wayne Booth's "reliable narrator" rule. The reliable narrator rule can be seen as an expectation or constraint with two parts: one, that the narrator will provide the reader with all data relevant to understanding the progress of the story; two, that the narrator's evaluation of that data will be truthful and helpful.[26] These sound, in fact, a great deal like Grice's maxims of relevance, quality, and quantity. As with Grice's maxims, Booth's principles are default expectations and may be violated in order to produce various effects in stories. Let us consider how an OT-type approach using Booth's rules can be applied to the activity of rereading in "The Garden of the Forking Paths" by Jorge Luis Borges. This analysis refers to my own experience reading the story.

The story is narrated by Yu Tsun, a Nazi spy. Here the constraints of my personal values interact with Booth's constraints on reliable narration; I cannot trust a Nazi spy. There is no textual reason to think Tsun would lie to me simply because he is a Nazi sympathizer; nevertheless, my negative judgment about his political ethics infects his narrative ethics. While at the beginning of the story, I cannot locate Tsun's unreliability specifically in his presentation of ir/relevant information or in his mis/evaluation of that information, my reading does reveal the disjointing of my interpretive experience from Tsun's. Suspicious, I reread everything he tells me over his shoulder, so to speak.

Tsun takes a desperate train ride into the country, closely pursued by a British inspector, to find the one man who can help him communicate the location of a British airstrip to Hitler's forces in Berlin. At the train stop he thinks is his, he asks some children on the platform if he is at Ashgrove, and they tell him yes. After he gets down from the train, the following scene ensues:

> A lamp lit the platform, but the children's faces remained in a shadow. One of them asked me: "Are you going to Dr. Stephen Albert's house?" Without waiting for my answer, another said: "The house is a good distance away but you won't get lost if you take the road to the left and bear to the left at every crossroad." I threw them a coin (my last), went down some stone steps and started along a deserted road. (93)

Children with shadowy faces who seem to know exactly what a complete stranger is looking for—this seems dangerous to me. Why else would Tsun have mentioned it, unless it were relevant, unless it were going to come back to haunt him later in the story? After all, I know my detective fiction. However, Tsun clearly evaluates the children as harmless and moves on without comment. In my newfound distrust of Tsun, I reevaluate the scene he has just presented me and decide the children are a threat to Tsun, as depicted in table 2.2. My decision shows me just what my problem with Tsun as an unreliable narrator is. Before, the two criteria, that a narrator must provide reliable access to relevant information and should reliably guide my evaluation of that information, were not ranked with respect to each other because they had not yet competed with each other in my reading experience. Now, however, they do compete, and my chosen interpretation of the scene with

Table 2.2 Decision about Tsun's Narration of Children

	Relevant info	Reliable eval
✓children a threat to Tsun		*
children not a threat	*!	

the children reveals that I rank relevance higher than evaluation in deciding whether or not to trust my narrator; the bold vertical line between the constraints indicates this crucial ranking. If I believed Tsun's assessment that the children were irrelevant to his mission in the long run, then I would have to admit he evaluated them reliably. But that would mean he took up my attention with totally irrelevant characters at a pivotal juncture in the story, and I cannot accept that in my current nervous state, wondering what Tsun's fate will be at the very climax of the story. Tsun has violated the Gricean maxim of relevance, leaving me no clues that would help me infer an ironic meaning from his irrelevant attention to the shadowy children; they thus loom in the background, unresolved and unrelated to any other action in the story. So I choose to believe, against Tsun, that the children are relevant, that Tsun has not recognized their threat to his story line, and I read on waiting for them to reappear from a dark alley at a crucial moment.

They do not. The story ends with no further reference to the children and no evidence of any effect they might have had on the outcome. If I had not done this exercise in examining the interaction of my expectations about Tsun's reliability, I would have remained unaware of a key element of my experience of Borges's storytelling. The children are not the only detail in Borges's story that seems to lead nowhere, that proves irrelevant to anticipating and understanding Tsun's mission. But the story is, in the end, about a maze, the "Garden of the Forking Paths." The irrelevancies in the narration I encountered are like wrong turns I took in the maze, leading to dead ends. These "wrong turns" create a powerful atmosphere of disorientation, frustration, and foreboding that could not be constructed another way. This exercise illustrates just one way in which an OT-type analysis can help explicate blockages or reinterpretations in the reading process and help the relative strengths of interpretive expectations to emerge at the same time.

Strengths and Weaknesses of OT

This interpretive decision about the reliability of a narrator is a very small move in the incredibly complex activity of reading. Taken together, all the reading expectations in play at any moment of reading (pragmatic, textual, personal, sociolinguistic, generic, etc.) describe the reader's current set toward or "filter" on meanings arising from the text. This is the most promising aspect of OT as applied to reading the hoaxes—the potential to recover from the immediate reader response to the hoaxes a ranked "filter" of expectations that both structures the individual reading process and provides a set of common reading expectations at the time (what I will refer to henceforth as a "common reading filter").

A few caveats are in order with respect to the adaptation of OT to reading hoaxes. First, there is the problem of history. In my reading of the Borges story, I had access to my own moment-by-moment interpretations during the reading process. The reader responses to the nineteenth-century hoaxes, however, are mostly post hoc evaluations of the hoaxes' truth or falsehood. So, while OT is capable of modeling the reading process online, as demonstrated above, the interpretive decisions modeled in this project will be after-the-fact evaluations, due to constraints in the historical data.

A second challenge in applying OT to reading the hoaxes is that OT as used in phonology assumes the universality of its constraints. As we have seen from the levels involved in reading, however, few if any can be universal. Yet some reading constraints are shared by reading communities with some common reading experience behind them, similar to Stanley Fish's "interpretive communities." While it is tempting to say that some reading expectations, especially the lower-level linguistic and pragmatic expectations, are shared by all readers/speakers of a language, that is a problem for future applications of OT to literary reading. This book will focus quite specifically on reader expectations of the genre of popular science news and of ethnoscience, or lay beliefs about science.

A few criticisms of my plan to apply OT to reading the hoaxes still remain to be answered, residing in the wake of a history of inappropriate or ultimately futile borrowings of linguistic methods to solve literary and rhetorical problems. Jacqueline Henkel points out some traditional pitfalls of these attempts.

> Critics may simply make an initially productive analogy account for literary facts too diverse; loosely applied at the outset, one notion (language as system, for example, or langue opposed to parole) applies to so many problems that it finally no longer provides enough resistance—a strong enough sense either of the goals of the original theory or of a coherent framework in which its various applications are articulated—to be usefully heuristic. The metaphor exhausts itself in local functions without pointing strongly toward an overall theory that will direct further critical practice. Or a (more strictly applied) metaphor starts as more genuinely parallel to a source concept (a literary rule analogous to a syntactic rule) but in a literary context accounts for so much more than was relevant in the linguistic theory that the literary version begins to collapse, and the literary facts the metaphor does not explain become increasingly obvious.[27]

I had to ask myself if my adaptation of OT was not subject to those criticisms and so to the same doom that has befallen generative text structure and the ideal reader. While I am still engaged in answering Henkel's challenges, I can offer the following signs of hope. First, OT can produce useful results outside of the analysis of hoaxes. It can model any reading activity that foregrounds readers' preconceptions about a genre—particularly parody and satire, in addition to stories such as "The Garden of the Forking Paths" that foreground the activity of reading. Finally, I have applied OT in the classroom as a teaching tool to show students how the criteria for what makes a good sci-fi movie or a good American president can interact and compete with each other in the construction of an evaluative argument. Visit my online tutorial (link listed below) for an example of this pedagogical technique.

Henkel's second challenge to borrowed methodologies is that they miss crucial and/or interesting literary questions because they are not methods founded on literary values and assumptions. In other words just as literary methods could not answer my questions about hoaxes because they were basically rhetorical and pragmatic questions, so my OT-based method could fail to answer literary questions. I found my method productive in addressing traditionally literary concerns such as the effects of history and politics on the reading experience, the effect of specific words and phrases on the reader, and even the sticky issue of authorial intention. That is because any application of OT relies on a method based in the field of application to derive the constraints that will drive the model. Since I had a primarily historical and rhetorical project, I used an established rhetorical method for analyzing historical acts of reading.

A related objection along the lines of Henkel's second challenge to borrowed methodologies could be that my OT-based method is not rigorous enough to be predictive—that is, to predict whether a hoax will be successful or not in advance of reader responses. OT as I have applied it is primarily historical and descriptive, not predictive. It starts with historical data about a particular reading act and a recorded decision and works backward to model the varying strength of a reader's expectations on that decision. The model is testable, of course, against the final decisions readers made. If the model fails to locate the deciding factors in their reading processes, to reflect all of the interpretive conflicts apparent in the archived responses, then it must be edited until an explanatory description is achieved. But the model itself does not predict if a particular hoax will or will not be successful. The reason for this is that the expectations that form the engine of the model are expectations about science news, not about hoaxes. A hoax is a parasitic genre that borrows its conventions from the genres it apes. Now, as an interesting indirect project, one could apply the model to a particular hoax to see if it satisfied readers' top expectations of a good science news story, but this would be a probabilistic prediction that would likely miss idiosyncratic factors in a hoax's success such as its serendipitous timing with other stories in the news or with the political climate at the time of the hoax's publication. OT works with historical judgments to open up the process of hoaxing for close examination. The conclusions that can be drawn from it pertain to the top-ranked reader expectations of science and why hoaxers chose to ridicule readers for these priorities.

Henkel does single out as promising ventures cooperative projects between language and linguistics that efface the privilege accorded to literary language and treat it as a coordinated communicative activity governed by conventions similar to Grice's. I believe that my application of optimality theory falls in line with those criteria. As a literary method, it also provides a way out of the Heisenbergian problem of fixing individual readings versus fixing context in the historical reading act: it provides a portrait of common reading codes in practice at a historical moment while still remaining capable of modeling one reader's reaction to a text. To put the case in linguistic terms, my OT-based method is able to account for both *la langue*, the communal possession of language, and *la parole*, the individual's creative practice of that communal possession. Thus, my methodology fits Andresen's desiderata for the future of linguistics in America as a discipline that "integrate(s) the study of language into use, society, history, and general cognition."[28]

So in conclusion, I have constructed a special method for answering my two pragmatic questions about the hoaxes—How did they fool readers? What social projects were the hoaxers trying to accomplish?—by first using rhetorical and new-historical techniques to recuperate the common expectations of nineteenth-century science newsreaders and then modeling their interaction and competition in the hoax-reading process via optimality theory. Further, I have argued that optimality theory shows promise as a functional rhetoric for reconciling linguistic and literary critical methods on appropriate projects. For readers interested in applying my method, I have posted a tutorial at http://www.nmt.edu/~lwalsh/OT_tutorial.html. I have also included an illustrative example and a glossary of key methodological terms at the back of this book for reference as we proceed to the discussion of the 11 major scientific media hoaxes comprising this study.

Chapter Three

Poe's Hoaxing and the Construction of Readerships

Edgar Allan Poe is the ideal figure with whom to begin any study of scientific hoaxing in America. Scientifically educated beyond many of his peers and a pioneer in at least two genres that showcase scientific epistemologies—science fiction and detective fiction—he embodies the tensions between the arts and sciences in the Jacksonian era. His hoaxes were public acts meant to call attention to these tensions, as they were written on science-related topics and carefully crafted and presented in popular news media for particular reading audiences. The later two, "The Facts in the Case of M. Valdemar" (1845) and "Von Kempelen and His Discovery" (1849), dealt with the psychological sciences, what are now deemed "pseudosciences": mesmerism and alchemy. But his first two hoaxes, "Hans Phaall—A Tale" (1835) and The Balloon-Hoax (1844), built fantastically intricate flying machines whose structure encapsulated a striking argument about reality, an argument Poe also makes in *Eureka*: that we should "put faith in dreams as the only realities."[1] Poe asked with his technological hoaxes: What was truer, or more real—that something actually existed and worked in the world, or that it could exist and work? Or, as Poe himself stated the case in his defense of M. Valdemar: "if the story was not true . . . it should have been."[2]

Poe developed this argument about reality through his hoaxing practices and the peculiar relationships with reality and readership that hoaxing enjoins. He and Richard Adams Locke, through the competition of their moon hoaxes in the Eastern media in 1835, innovated the genre of the scientific media hoax in America. Both writers baited their hooks with a cluster of rhetorical lures that mimicked the popular science reports their readers were accustomed to encountering in almost every newspaper and magazine. Poe, particularly, was explicit about what he thought his readers expected from a science report; he discussed these reader expectations several times in different formats dating from the overshadowing of his first hoax by Locke's hoax. By intuiting and then reperforming his readers' expectations about how science was read, Poe managed to hoax a good percentage of his readers in at least two of his four attempts. He did this for several purposes: first, to demonstrate his creative authority over his readers even to the point of altering their realities; second, to criticize those readers for their admiration and funding of professional scientists instead of professional artists; third, to reveal the vulnerability of purely inductive, Baconian science and thereby lay the groundwork for his own imaginative epistemology, outlined in *Eureka*; and fourth, to materialize a community of fellow geniuses sympathetic to *Eureka*'s epistemology.

Because of his explicit attention to what makes a hoax work, Poe's hoaxing practices offer an ideal opportunity to test the methodology laid out in the previous chapter—using contemporary reader responses to elicit and structure a "filter" of reading expectations that Poe's readers might have held in common when coming to his hoaxes. First, however, it is important to establish how Poe learned science and the conventions of writing it for a lay audience. After that, we will examine the competition between Poe's Hans Phaall and Locke's Moon Hoax in the media and the various reactions to the hoaxes in order to glean readers' expectations of a "true" popular science report; I will also examine the rhetoric of the popular science article at the time, a source of reader expectations about the genre. Using optimality theory, I will model how these expectations interacted and competed with each other in producing either belief or doubt in the moon hoaxes. Then I will extend the method to examine the rest of Poe's hoaxes and, based on reader reactions to them, to synthesize a common reading filter of antebellum reader expectations for science news. After a discussion of the ways in which a rhetorical methodology solves problems that have plagued Poe hoaxing scholarship—particularly problems with understanding his choice of hoaxing when constructing relationships with his public, I will conclude by connecting Poe's hoaxing to his scientific epistemology in *Eureka* and suggesting that both projects reveal Poe gesturing toward community.

Overview of Poe's Scientific and Rhetorical Acculturation

This section considers how Poe came to know science as a cultural practice. According to Roland Barthes, who analyzed the "structuration" of a Poe hoax in depth, Poe internalized, through learning to read and write about science, a "*cultural code* . . . the code of knowledge, or rather of human knowledge, public opinion, of culture as transmitted through books, education, and in a more general, more diffuse way, through all sociality."[3] We will return to Barthes's particular structuring of this notion of code after examining the media through which Poe acclimated to scientific culture and its popular rhetoric.

Certainly, Poe's excellent primary education played a crucial role in the development of his scientific rhetoric, particularly in terms of how to cope with an audience. Under the aegis of his foster father, John Allan, Poe was educated in excellent private schools both in England and in America. From the age of about seven to the age of eleven, Poe studied French, Latin, history, and literature at the Manor House school in Stoke-Newington outside London.[4] When his family returned to the United States in 1820 after a business venture of Mr. Allan fell through, Poe was enrolled in Joseph Clarke's private school in Richmond. He studied more classical languages there; from tuition bills and letters from Clarke, we know Poe was reading Horace's *Odes* and Cicero's *De Officiis* (and likely *De Oratore*) in Latin and Homer in Greek. By the age of sixteen, Poe was fluent in French, dexterous in Latin, and triumphant in speech competitions with his classmates.[5] While reading Cicero and copying and "capping" Latin verses likely amounted to his only formal rhetorical training at this point, his studies undoubtedly acquainted him with the classical structure of arguments and techniques for persuading audiences. The sciences, even natural philosophy or theology, were not part of a traditional primary education at this time. They were more advanced studies reserved for the university.

Poe entered the University of Virginia in 1826, and though he got himself kicked out for gambling by December of that year, he nonetheless distinguished himself in his course of modern and classical languages.[6] His library card reveals that he checked out many history texts in addition to works by Voltaire and Byron. Although Poe was not enrolled in any rhetoric classes and left behind no evidence of having checked out or bought any rhetoric texts, he was active in a debating society at the university, which indicates that his interests in argumentation and persuasion had not waned.[7] Susan Booker Welsh in her dissertation "Edgar Allan Poe and the Rhetoric of Science" argues that George Campbell's *Philosophy of Rhetoric* was the primary rhetoric text in American colleges from 1800 through 1850 and that Poe must have read it. Although there is no direct evidence of this, Welsh cites as circumstantial evidence Poe's review of Leigh Hunt's *Imagination and Fancy.* The review seems very Campbellian in overall tenor, as it criticizes Hunt for his purely inductive model that excludes speculation, which Poe believed was the duty of literary philosophers as public truth makers.[8] Whatever his actual indebtedness to Campbell might be, Poe openly employed many of the principles of faculty psychology in his Dupin tales and in his *Philosophy of Composition.*[9] And, famously, Poe touts the "faculties" of intuition and the imagination over syllogistic logic in *Eureka*.[10] Rather more crucial for the purposes of explaining Poe's hoaxing practices is the possible Campbellian legacy of a stochastic model of belief. Poe recognized that belief, for his readers, was not a matter of positive demonstration but a matter of likelihood. This may explain his repeated emphasis, in his rhetorical analysis of Locke's Moon Hoax, on scientific detail as the most important factor in making a hoax seem probable, and therefore acceptable, to the reader.

Although Poe's formal education ended with his withdrawal from the University of Virginia, Poe still read widely, especially in science, which monopolized his reading and writing attentions even from this early period. He published *Al Aaraaf, Tamerlane and Minor Poems* in 1829, and the first poem in the collection was the "Sonnet—To Science." Traditionally viewed as an early expression of Poe's antipathy toward contemporary Baconian inductive science, there is resident in the poem, however, an inkling of the fascination with science—especially mathematics, astronomy, mechanics, cryptography, and psychology—that would dictate Poe's choice of topics for the rest of his writing career. In the sonnet Poe concludes that science has "torn from me/The summer dream beneath the tamarind tree."[11] While this image amplifies the theme developed in the rest of the poem of the damage science has done to the arts, another meaning lies very close to the surface: Poe cannot leave science alone. It enthralls him; it disturbs his rest. One might anticipate from this very early sentiment what Poe indeed goes on to attempt in *Eureka*—a reconciliation between science and imagination, the two forces that lay claim to Poe's intellectual life.

Rather than continuing along this vein, however, and winding up psychoanalyzing Poe's predilections for science and technology, this project, due to its focus on conventions of reading popular science, requires instead an examination of the scientific climate in which Poe read and wrote in the 1830s and 1840s. In chapter 1 we got a sketch of this climate, essentially a vortex of desire between professionalizing scientists who wanted money and the lay public who wanted technologies and certainty. Poe's hoaxes addressed the lacuna at the center of this vortex by exploiting three specific social movements that were trying to erase or "bridge" it: "scientific" spectacles like those in Barnum's American Museum, mesmerism and other "pseudosciences," and scientific treatises and articles written for general audiences.

Science as Spectacle in Antebellum America

As Judith Yaross Lee has summarized the situation, "the role of the lay public in nineteenth-century science shifted from participant to spectator."[12] Spectacle was as much a part of science during this period as research was. Swiss naturalist Louis Agassiz attracted much of his funding through his impressive collections of exotic stuffed species of animals from all over the world, and he was not the only scientist to walk wide-eyed prospective investors down aisles of stuffed peacocks and tortoises.[13] It is little wonder, then, that P. T. Barnum's collection of freaks of nature in his American Museum in Boston could flourish during the middle of the century and, yet, still be counted "scientific" by a distinguished natural scientist such as Spencer Baird, himself a veteran peacock stuffer.[14]

Barnum himself was only exploiting a tendency in America at that time to believe in natural wonders, a "predisposition to accept the mechanically probable or the organically possible . . . [that] was a peculiarly patriotic position in Jacksonian America."[15] There were legitimate reasons for Americans to put their faith in the seemingly fantastic. The American subcontinent was being actively explored by expeditions such as Lewis and Clark's and Wilkes's (the United States Exploring Expedition). These adventurers returned with an astounding bounty of natural specimens along with Native American cultural artifacts. The first bison and gila monsters appeared just as outlandish to New Yorkers as Barnum's Feejee Mermaid (cobbled together from a mummified monkey and a fish). In a way the feeling of many Americans toward new scientific discoveries was like Poe's toward his flying machines: an optimistic focus beyond what was to what could be. Barnum thus capitalized on a nearly inexhaustible supply of cheerful, patriotic credulity on the part of Americans who were as agape on their new home continent as children in a candy store. To illustrate this good-natured naïveté, cultural scholar Jonathan Elmer discusses the sign in Barnum's American Museum that pointed "To the Egress." When patrons, curious to see a female "egre," would pass through the door, they would find themselves out in the alley with no way back into the museum but to pay the fee again. Elmer claims that being duped provided no small amount of pleasure for those whose egos were not caught up in being right all the time and that Americans, in general, loved to be fooled. "What Barnum [sold], by means of his objects, [was] interpretation."[16] It was the experience of perceiving and deciding that Americans enjoyed, the freedom they felt in the Jacksonian era to make the reality in which they wanted to live.

Poe clearly exploited the same cultural dynamics that Barnum did when writing his hoaxes about balloon trips to the moon, Atlantic-crossing balloons, and suspended animation. When Poe sniped in his criticism of Locke's Moon Hoax that newsreaders valued sensation over hard fact, he feigned disdain for their comfort with ambiguous truth-values. However, as evidenced by his arguments for the value of his stories on the basis of their plausibility and potentiality, not their factuality, Poe clearly shared some of Locke's readers'—and Barnum's customers'—love of suspended meanings and multiple possible realities.

"Pseudoscience" in Antebellum America

This conflation of science and spectacle also created a space for play in the public sector for what are now considered pseudosciences but what were then becoming known as "social sciences": magnetism (mesmerism) and phrenology in particular. The public life of mesmerism is of the greatest interest to us, since Poe used it as the topic of his hoax, "The Facts in the Case of M. Valdemar." A more detailed discussion of Poe's education in this field appears in the analysis of that hoax later in this chapter. However, a few comments are pertinent here. One important observation, given our anachronistic perspective, was that mesmerism was one of the best-reputed of the "pseudosciences" because the hypnotic trance was a real phenomenon and mesmerists' claims about the myriad effects of the body's magnetic fields were unfalsifiable at that time in medical history; also, mesmerism had no serious competitors in terms of explanatory theories of the subconscious. Consequently, the *American Journal of Science*, Benjamin Silliman's well-respected Yale general science journal regularly featured articles on mesmerism along with more standard reports of discoveries in chemistry, geology, and physics.

Mesmerism, like medicine during this time, was a science with many lay practitioners. Therefore, it helped create a sense of continuity between American public life and the increasingly rarified communities of science. In addition to this liaison, mesmerism forged a second connection. Pseudosciences such as mesmerism earned a great deal of their credibility via their humanitarian aims. Their practitioners preached them as efficacious for the improvement of human relations and living conditions, matters of real concern to antebellum Americans confronted daily with the social problems of slavery and of industrialization. Allying itself with the other budding "social sciences" of psychology, feminism, and sociology, mesmerism aimed at social reform.[17] It applied scientific principles to the prediction of behavior; it offered an illusion of some kind of control over the bewildering array of motivations and styles of personal interaction that were coming into contact (and conflict) with each other during the population explosion of the Jacksonian era. With this innate appeal of social control working to its advantage, mesmerism and "other equally delicious ism[s]," as Poe deemed them while lampooning their proliferation in *Eureka*,[18] constituted a serious attempt at reconciling the explanatory power of the increasingly Brahministic academic sciences with the social concerns of the average working American. Acknowledging the bridging function of practices such as mesmerism is vital to understanding Poe's hoaxing. Ever a lover of the liminal, Poe recognized the power of the "isms" to tap into the sympathies of his readers; thus, he chose mesmerism and alchemy as engines ideally suited to driving home the effect of his last two hoaxes, "The Facts in the Case of M. Valdemar," and "Von Kempelen and His Discovery."

Popular Science News in Antebellum America

Poe also availed himself of the last "bridge" to appear between popular and elite scientific culture—the rapidly increasing number of pages in the literary weeklies and monthlies (and, beginning in 1835, the penny press) devoted to science and technology news. The innovation of the penny press merits deeper discussion in the following section with respect to Locke's Moon Hoax, but Poe's other periodical science reading can be surveyed briefly here. Poe clearly read Silliman's *Journal*, but he also demonstrated familiarity with the science news in the *Home Journal* and Evert Duykinck's *Literary World*, both magazines meant for the general reader.[19] Special "general knowledge" magazines had even developed during Poe's lifetime for instructional use in the home. The *Family Magazine*, founded in 1833, published articles on the geology of earthquakes and volcanoes next to poetry for the family to read together. Editors clearly recognized that the public thirst for science in the media could not longer be satisfied by the rapidly specializing scientific communities, so a popular genre of "science writing" gradually developed, with some papers and magazines actually beginning to retain a science writer on staff. According to Carolyn D. Hay's study of the founding of the National Association of Science Writers, antebellum newspapers covered a range of scientific topics mirrored by the advertisements mentioned earlier: "medicine, agriculture, inventions and technology, pure science, exploration, aviation ," with technology receiving by far the most press.[20]

Scientists themselves realized the popular press was stepping in where they had stepped out, and some were troubled by the resulting misinformation of the public. Joseph Henry complained to a friend, "In this country, our newspapers are filled with the puffs of quackery and every man who can burn phosphorous in oxygen and exhibit a few experiments to a class of young ladies is called a man of science."[21] But even if they felt some responsibility for the "quackery" running rampant in the papers, few scientists were willing to write for the public on a regular basis.[22] It required time they simply did not feel they had. A few scientists perhaps even enjoyed treading the fine line that separated scientific fact from fantasy in the public gaze. It is a persistent rumor, for instance, that French astronomer Jean-Nicolas Nicollet helped Richard Adams Locke craft the details of his Moon Hoax in 1835. After the Civil War, Dr. William Osler, professor of medicine at Johns Hopkins, entertained himself by sending hoax medical reports, usually sexual in tenor, to prestigious medical journals through his alter ego Dr. Egerton Yorrick Davis.[23] Though media hoaxes authored by scientists have been rare occurrences throughout the twentieth century and up to the present, the Sokal hoax, which will be treated in the last chapter of this book, is a notable exception.

This brief survey of the scientific life of America in the Jacksonian period leaves us with the sense that even if Poe had restricted his scientific curiosity as a reader entirely to the popular press and the court of "public opinion" as Barthes would term it—Lyceum science lectures, advertisements and news about mesmerism and other scientific wonders, and the daily innovations in transportation and communication—he would have had an impressive scientific education. However, Poe further sought out and eagerly devoured the writings of European scientists in American reprint. Especially interesting, both for the purposes of considering his hoaxes and for appreciating the breadth of Poe's scientific curiosity, are Poe's readings of the works of astronomer Sir John Herschel and chemist Sir Humphrey Davy, the journals of the balloonist Monck Mason, the mathematics of Pierre Laplace, and the travel narratives and cosmology of Alexander Humboldt. Poe also consulted older works by Kepler, Newton, and Bacon in writing *Eureka*. Moreover, he armed himself with the natural philosophy of the German and English romantic schools, including Kant and Byron and, most powerfully, the rhetoric of Coleridge in the *Biographia Literaria*, which figured centrally in Poe's *Philosophy of Composition* and the *Rationale of Verse*.[24] From these last works he inherited the war between science and art that he would struggle to mediate in his literary works, beginning with "A Sonnet—To Science," continuing through his hoaxing, and culminating in *Eureka*.

The influences of these scientific and technological authors and others will be examined more closely as each of Poe's hoaxes is considered in turn. But even a sampling impresses the reader with the resources available to someone like Poe, who had a good education, a small amount of money, and wished to learn about science in the 1830s and 1840s. However, neither money, nor breeding, nor education guaranteed a working knowledge of the most basic of scientific principles, or so lamented an essay by C. L. Barritt in the 22 February 1845 edition of the Poe-edited *Broadway Journal* entitled "Why Are Not the Sciences Better Understood?" Barritt complains in his essay that the cultured "young gentlemen" of the day did not even know "why they are warmer in a woolen blanket than in a cotton one of equal weight . . . or why a white hat is cooler than a black one."[25] He lays some of the blame for this regrettable state of scientific ignorance on the American primary education system; however, Barritt more squarely indicts the characters of these gentlemen, namely, their laziness and fondness for the opera, fashion, and the romance novel over a little serious reading that would be "of more credit of the person, than being able to correct a false step at a cotillion."[26] Poe had certainly had an upper-class acculturation through John Allan's family. But whether because of that upbringing or in spite of it—since after he lost the post John Allan got for him at West Point, he had almost nothing further to do with the Allans—Poe seemed to take his own science education very seriously. As evidenced by his reading and, in his early writing, by the copious scientific details of observation and navigation weighing down even such an unscientific story as "MS Found in a Bottle," Poe was committed to the weird hybridization required in the writing of science between text on the one hand, and the antitextuality of immediate sensory perception and measurement on the other. Indeed, if "MS Found in a Bottle" does not in fact begin autobiographically, it could have, as Poe would say:

> Of my country and of my family I have little to say. Ill usage and length of years have driven me from the one, and estranged me from the other. Hereditary wealth afforded me an education of no common order, and a contemplative turn of mind enabled me to methodize the stores which early study very diligently garnered up.—Beyond all things, the study of the German moralists gave me great delight; not from any ill-advised admiration of their eloquent madness, but from the ease with which my habits of rigid thought enabled me to detect their falsities . . . Indeed, a strong relish for physical philosophy has, I fear, tinctured my mind with a very common error of this age—I mean the habit of referring occurrences, even the least susceptible of such reference, to the principles of that science.[27]

The story goes on to relate a horrible sea adventure both very supernatural and unscientific. But Poe remained committed throughout his writings to finding that place where "physical philosophy" broke down and imagination took over, and marking that spot with words that partook of the traditions of both science and art. Poe was certainly not unique for his time in his dedication to both the arts and the sciences: Emerson's study of natural history has been detailed in depth by Lee Rust Brown; Taylor Stoehr has documented Hawthorne's fascination with nascent social sciences like phrenology; and Melville's preoccupations with industrial and marine science and technology are manifest in stories such as *Moby Dick* and "The Paradise of Bachelors and the Tartarus of Maids." However, Poe was unique in making a failure envelope—the line at which the stress between art and science became too much, where words failed to describe experience, experience failed to adhere to scientific principle, and words and science and experience all failed to reliably yield what was true—as the guiding arc for his thinking and writing life.

Neither was Poe content to wrestle with these fractures alone. Through the hoax he coerced his readers into experiencing those problems in a unique way. Poe lashed out at his readers through his hoaxes for creating a society he wanted to succeed in and could not, a society that made money its end-all-be-all and valued "dull realities," as he expressed it in "Sonnet—To Science," over what could be.[28] It galled Poe that he could barely make ends meet, that publishers and booksellers routinely abused writers like him while scientists (from his perspective) raked in government and private funding for their technological and medical inventions.[29] To those readers he managed to dupe with his hoaxes, then, Poe communicated for at least a few moments the discomfort he felt living in the "reality" of an America committed to what science could do and buy, not the possibilities science afforded the imagination for apprehending the true core of the world. This is a powerful motivation, indeed, for choosing the hoax as a means of criticizing the ascendancy of professionalized Baconian science in America.

The Contest between Hans Phaall and Locke's Moon Hoax: Revealing Reader Expectations

Poe's first attempt at a hoax is a confusing one to begin with, because it is uncertain if Poe even meant "Hans Phaall—A Tale" to be taken as a news report of a journey to the moon when it came out in the June 1835 edition of the *Southern Literary Messenger*, which he was then editing for owner Thomas H. White. Later, Poe would claim that it was both a "hoax" and a "jeu d'esprit," both that it was meant to fool its readers, and that it could not have, given its "tone of mere banter."[30] The story concerns a burgher of Rotterdam, one Hans Phaall, who constructs a balloon and sails to the moon in order to escape creditors. Along the way, Phaall pioneers an air compressor to help him breathe in space and experiments with the reactions of a cat to the vacuum between the earth and the moon. The moon itself is apparently inhabited by dwarflike people, or so reports a letter flown back to Rotterdam from the moon in the same balloon four years later. Poe clearly intended to continue the tale, elaborating on the moon inhabitants, but Locke's Moon Hoax stole his thunder, as will be seen shortly. While much of the language of Hans Phaall is little short of goofy, in keeping with its original subtitle, A Tale, the opening of Poe's first attempt at a hoax sounds newsy enough:

> By late accounts from Rotterdam that city seems to be in a singularly high state of philosophical excitement. Indeed phenomena have there occurred of a nature so completely unexpected, so entirely novel, so utterly at variance with pre-conceived opinions, as to leave no doubt on my mind that long ere this all Europe is in an uproar, all Physics in a ferment, all Dynamics and Astronomy together by the ears.[31]

As will be apparent when we examine the form of science reports from this historical period more carefully, this opening actually conforms remarkably well to reader expectations of true science discoveries. However, at least four things in addition to the title of the story were already working against Poe if he expected Phaall to be taken seriously as an aeronaut. The tone of "together by the ears" is off, and Poe has given his byline to the story, a practice more typical of literary than news writing at this juncture in the history of American journalism. In addition, Poe's literary reputation among southern readers at this point was shaped largely by the award-winning "MS Found in a Bottle," a horrifying and obviously fanciful tale of a phantom voyage, published in the Baltimore *Visiter* in 1833. Finally, the editorial introduction to this issue of the *SLM* by Edward V. Sparhawk claims that Hans Phaall "will add much to [Poe's] reputation as an imaginative writer" even as he notes out of the other side of his quill that in the days of frequent and well-publicized experiments in balloon aviation, "a journey to the moon may not be considered a matter of mere moonshine."[32]

As might be expected after this caviling introduction, there was no serious debate over the truth of Hans Phaall. Dwight Thomas and David Jackson in *The Poe Log* list nine notices of the story, and all of them focus on the humor of the piece, not its possibility. The Charleston *Daily Courier* praised "the minuteness of detail, which properly belongs to truth" in the story but went on to deem it "one of the most exquisite specimens of blended humor and science that we have ever perused."[33] Poe mentions two reviews in a letter to Thomas White that focus on the opening as the weakness of the piece.[34] Indeed, as the story unfolds, it sounded very little like a news story. As Phaall's strange makeshift balloon descends from the sky over Rotterdam, Poe frames the Dutch response in a manner that contrasts sharply with both the relatively matter-of-fact language of the first paragraph of the story and with the parade of scientific minutiae about Phaall's balloon a few columns later:

> What could it be? In the name of all the vrows and devils in Rotterdam, what could it possibly portend? No one knew—no one could imagine—no one, not even the burgomaster Mynheer Superbus Von Underduk, had the slightest clue by which to unravel the mystery: so, as nothing more reasonable could be done, every one to a man replaced his pipe carefully in the left corner of his mouth, and, cocking up his right eye towards the phenomenon, puffed, paused, waddled about, and grunted significantly—then waddled back, grunted, paused, and finally—puffed again."[35]

Strange language indeed for a news report. Poe acknowledged his critics' negative reaction to the "tone of mere banter" in the tale; he believed, in fact, that the humorous tone of Hans Phaall was the principal reason it did not fly as a hoax.[36] Accordingly, he tuned the language of his future hoaxes to a more newsy resonance. However, in spite of its initial fizzling, the public career of Hans Phaall was just beginning. It got a kick start from another fantastical moon story that appeared just two months later in the brand new penny daily, the New York *Sun*.

Before the second moon hoax is considered, a digression is in order to explain the place of the *Sun* in the reading life of antebellum New York. Founded in 1833 by Benjamin Day, the *Sun* was the harbinger of what would be termed the "penny press"; it was a single sheet folded to four pages that sold for a penny, containing, in addition to the usual copious advertisements, news items appealing especially to the working class and new immigrants, its target demographic.[37] The *Sun* contained notices of all the sorts of things that still make up the bulk of conversations at coffee shops and street corners: fires, accidents, the daily police blotter, spectacles and scientific wonders on display at local museums, and even humorous recitals of domestic disputes and other human interest stories. These quotidian topics actually constituted a radical departure from the reading material New Yorkers previously had at their disposal. The literary weeklies, which cost six cents instead of a penny, were almost all owned by political organizations that filled their pages with political news and party propaganda. The penny paper's low price and gossipy material aligned it with the Jacksonian democratic spirit in general and with the working class and immigrants in particular. New York *Herald* editor James Gordon Bennett trumpeted in the pages of his penny daily: "I feel myself in this land to be engaged in a great cause—the cause of truth, public faith, and science against falsehood, fraud, and ignorance."[38] He had some justification for his claim to being the paper of "public faith" because of the sheer number of people who read the penny dailies. Frank Luther Mott argues in *American Journalism: A History* that the 1830s saw more American newsreaders than ever with an influx of immigrants who took their voting rights seriously and read the papers for political information; with public education squashing illiteracy to 9 percent or less and creating an ever-growing percentage of women readers;[39] and, with advancements in the material conditions of reading as simple as better lighting in homes and streets.[40] So the penny press had a broad-based lower- and middle-class audience whose numbers were interested in information both for entertainment and for political use.

The New York *Sun* had on staff a science writer, one of the few at the time, named Richard Adams Locke. Benjamin Day had hired him to cover the trail of the infamous cult leader Matthias the Prophet. Locke's skillful handling of the bizarre metaphysical and religious elements of the story put the *Sun* ahead of its rival penny dailies, Bennett's *Herald* and the New York

Transcript, in subscriptions. So after that Locke was employed to cover topics of general interest in the sciences and philosophy especially for the *Sun*.[41]

On 21 August 1835, a blurb appeared on page two of the *Sun* claiming that Sir John Herschel, who was known to most news readers to be engaged in observations in South Africa with his new telescope, had made some remarkable discoveries. There was nothing further for four days, and then, on the front page of the 25 August *Sun*, after an editorial note advising readers that the following story was reprinted from the *Edinburgh Journal of Science*, minus most of the "more abstruse and mathematical parts" of the discovery, the story commenced:

> In this unusual addition to our *Journal*, we have the happiness of making known to the British public, and thence to the whole civilized world, recent discoveries in Astronomy which will build an imperishable monument to the age in which we live, and confer upon the present generation of the human race a proud distinction through all future time. It has been poetically said, that the stars of heaven are the hereditary regalia of man, as the intellectual sovereign of the animal creation. He may now fold the Zodiack around him with a loftier consciousness of his mental supremacy.[42]

After some preliminary reflections on the wonders of astronomy, the story went on to detail how the lens for Herschel's telescope was ground, how it was transported to South Africa and the observatory assembled, how the problem of insufficient lighting was overcome, and so on, to an almost overwhelmingly technical degree. The next day the *Sun* ran the columns revealing what readers had been waiting for: the moon bison, man bats, moon poppies, and moon beavers that Herschel had glimpsed through his telescope. The paper sold 19,360 copies, the largest circulation of any paper ever in America.[43] The series went on for a week, and the reaction of the competing media was evenly divided. Of the major New York newspapers surveyed by both me and by Ormond Seavey in his 1975 edition of Locke's story, seven initially supported the Herschel report, two stayed on the fence, and seven criticized it, although many of the negative responses were indirect or ironic, perhaps hedging just in case a confirmation from Herschel came through. Two of the critical papers, the *Journal of Commerce* and the *Herald*, suggested that the whole thing was a hoax; the *Journal of Commerce* even connected Richard Adams Locke's name with the story on the strength of a "confession" by Locke to a *Commerce* reporter in a pub. James Gordon Bennett attacked Locke's character in a *Herald* editorial, intimating aristocratic dissolution involving a chambermaid. Locke's only public response to any aspect of the

Figure 3.1 Richard Adams Locke, author of the Moon Hoax. *Courtesy, Library of Congress*

brouhaha was to take the moral high ground in a defense of his good name.[44] Bennett and the other editors then proceeded with a more indirect tone for the duration of the two-week media debate following the appearance of the story. A rumor cropped up that scientists from Yale had taken a train up to New York to consult with Locke about the story but had to return unsatisfied; the *Sun* never admitted to the hoax.[45]

Whatever people may have finally decided about the Moon Hoax, it was the first major hoax of its kind,[46] and it fooled most New Yorkers at

Figure 3.2 J. W. F. Herschel by Julia Cameron, 1858. *Courtesy, Wilson Centre for Photography*

first—including educated people such as clergy and scientists—according to all extant commentary by Locke's contemporaries.[47] One of these, British writer Harriet Martineau, said she encountered on her 1835 visit to the northeastern states an atmosphere of almost complete credulity surrounding the hoax, with very few exceptions. She defended American readers, however, by claiming that the Moon Hoax would have gone even further in Europe, since she deemed the quality of science education in American comparatively high.[48]

People either believed or suspected Locke's story; parodies immediately flourished in the *Herald* and elsewhere, but nothing was certain except for the *Sun*'s subscription figures. Locke's unwitting accomplice in the hoax, Sir John Herschel, did not find out about the hoax until four months later, and his reaction evinced an appreciation of the human imagination that even Poe would have approved: Herschel claimed in a letter to the American captain who brought him a clipping of the hoax that it was a "perpetual reminder how trivial are the discoveries which all our boasted science has yet revealed or is [likely?] to reveal for ages to come in comparison of what exists unknown and unsuspected among the realities of nature."[49] Locke himself claimed ex post facto that he had meant his story as a satire of the famous astronomer Thomas K. Dick's bizarre plan to communicate with moon beings through geometric shapes; however, in view of how the report was taken, Locke felt it was an "abortive satire . . . and in either case I am the best self-hoaxed man in the whole community."[50]

Edgar Allan Poe made up his mind almost immediately about the hoax. In a letter to editor John Kennedy dated less than two weeks after the Moon Hoax finished its run in the *Sun*, Poe accused Locke of plagiarism:

> Have you seen the "Discoveries in the Moon"? Do you not think it altogether suggested by *Hans Phaall*? It is very singular,—but when I first purposed writing a Tale concerning the Moon, the idea of *Telescopic* discoveries suggested itself to me—but I afterwards abandoned it. I had however spoken of it freely, & from many little incidents & apparently trivial remarks in those *Discoveries* I am convinced that the idea was stolen from myself.[51]

Apparently Poe aired these charges publicly in the papers but did not pursue them legally, and he was not the only one who saw a similarity between the two moon stories; the New York *Transcript* printed them together and speculated they were by the same person.[52] Poe eventually had to relinquish his literary vendetta against Locke as it proved unproductive.[53] However, in the process of protesting both the similarities and the differences between the two hoaxes, Poe offered an insightful critique of readers' assumptions about popular science news. His reaction will be one of those we consider as we try to account for how the Moon Hoax fooled its readers.

Collecting Reader Expectations

It is impossible to determine precisely how readers in the 1830s and 1840s read popular science reports. No judgments made from our perspective are accurate or comprehensive, and contemporary opinions are biased by their imbrication in the reading culture. As James Machor describes the difficulty, "[T]he impossibility of full and unmediated access to historical readers ineluctably limits efforts to 'recapture' reading as a historical act."[54] This is precisely the problem that has dogged historical reading researchers, as discussed in chapter 2. However, the problem cannot be sidestepped in studying historical hoaxes, as readings of them without reference to their original contexts of publication cannot account for their rhetorical effects. Roland Barthes's influential reading of Poe's hoax "The Facts in the Case of M. Valdemar" is an excellent example of the methodological pitfalls of anachronistic reading. Barthes's study begins by claiming history is unnecessary for the reading of M. Valdemar, but he cannot avoid invoking it to explain how the story taps into a roiling nineteenth-century fervor over mesmerism.[55] When it comes to interpreting the cultural codes the story supposedly carries, Barthes eschews the publication history of M. Valdemar, including the crucial fact that it was originally read not as science fiction, as Barthes reads it, but as *news*. That basic assumption about the truth-value of the piece radically alters the key for all codes in M. Valdemar, but with his insistence on context-free reading, Barthes cannot acknowledge this mutation ineluctably introduced into his analysis nor gauge how it affects his results.

This text is an attempt to find a corrective to these kinds of critical problems. Given that any study of the rhetorical effect of hoaxes must work within the context of their immediate reception, this book will treat the hoaxes as case studies in the identification and reperformance of expectations that readers of popular science in the 1830s and 1840s might have had. The recuperation and ranking of these expectations will provide a portrait of science newsreading at different times in the nineteenth century, a common reading filter that accounts for the historical context surrounding individual readings while remaining capable of describing the readings themselves.

Reading expectations naturally range over a broad field, from expectations about particular authors' writing styles to expectations completely beyond the pale of study—like a personal predisposition to disbelieve anything anyone says because of a recent betrayal by a friend. For the purposes of this book, I will focus on recuperating only expectations concerning the genre of the popular science report and ethnoscience—what Barthes would call the "code" of science. As we analyze the Moon Hoax, I will use the method described in chapter 2 to identify repeated topoi in archived debates over the veracity of the hoax. I will validate these collected expectations against three other sources: Poe's hypotheses about the assumptions made by average readers of Hans Phaall and the Moon Hoax, a parody of the Moon Hoax in the New York *Herald*, and a sample of popular science articles from the time period.

My reasons for using these supplements to the primary reader reactions are all based on a critical assumption: a genre is a ritualized communicative activity. It is a recognizable pattern of rhetorical moves that solidifies as writers and readers repeatedly engage each other to accomplish a particular communicative purpose. Therefore, the formation of a genre is dynamic and circular. The participants' expectations form the genre as the genre forms their expectations. From this perspective, it is reasonable to look to the patterns within the genre itself to understand how readers learned what to expect from it. Further, parodies of the genre are immensely useful for identifying key reader expectations since parodies' humorous effects depend on their ability to "do a good impression" of exactly those generic features most salient to readers. Finally, in the case of successful hoaxes such as the Moon Hoax, their writers' guesses about their readers' expectations and priorities are also a useful data source—after all, these guesses were tested on real readers, and at least some of them worked. In other words, we know that Poe and Locke correctly guessed a percentage of their readers' expectations about popular science news because the Moon Hoax, in particular, succeeded in fooling a great number of those readers into thinking it was a real science report. It was in toto a successful experiment in producing a certain reader response via a series of certain rhetorical moves. All of these sources furnish evidence of reader expectations that we will reconcile with the expectations recollected from reader responses to the hoaxes; these expectations will form a theoretical foundation for examining Poe's future hoaxing practices, as well as those of the hoaxers that followed his lead.

Reacting to the Moon Hoax

Readers who paid a penny to read the Moon Hoax and thirteen cents to buy the elaborate lithographs the *Sun* printed up of its moon bison and man bats were not the only New Yorkers having to decide for themselves if the story were true. All of the *Sun*'s rivals in the popular press felt the pressure generated by the paper's astounding circulation figures; quickly, they had to decide whether or not to jump on the lunar bandwagon. So, in addition to the six responses from individual readers that Ormond Seavey collected for his 1975 edition of the hoax, we have the reactions of sixteen editors justifying their decisions either to reprint the story as true or decry it as a hoax. In all of these discussions, the same topoi come up over and over again, signaling the readers' expectations of what a true popular science report should read like. In Table 3.1 I have summarized these topoi (including the collocative series indexing them) and expectations with illustrative examples from some of the responses. The complete set of responses can be found in appendix B at the end of this book.

Table 3.1 Expectations of Popular Science News Collected from Reader Responses to "Moon Hoax"

Topos (Collocative Series)	Expectation	Examples
Entertainment (entertainment, pleasure, ingenious, story, fable, etc.)	"You read popular science news primarily for entertainment, not to make decisions affecting your world view."	"We . . . commend the article to our readers as one from which they will derive much **entertainment** whether they wholly believe it or not." "In sober truth, if this account is true, it is most enormously wonderful, and if it is a **fable**, the manner of its relation, with all its scientific details, names of persons employed, and the beauty of its glowing descriptions, will give this **ingenious** history a place with 'Gulliver's Travels' and 'Robinson Crusoe.'"
Foreign (transatlantic, Europe, Edinburgh, Scotland, Capestown)	"Anything foreign is good and probably true."	"We think we can trace in it the marks of **transatlantic** origin." "As it professed to be a republication from the **Edinburgh** Journal of Science, it was some time before many persons, except professors of natural philosophy, thought of doubting its truth."
Sensation (sensation, intense interest, astonishment, importance, marvelous, wonderful)	"Sensational elements in a story have a high literary and truth-value."	"GREAT ASTRONOMICAL DISCOVERIES!—By the late arrivals from England there has been received in this country a supplement to the Edinburgh Journal of Science containing intelligence of the most **astounding interest** from Prof. Herschel's observatory at the Cape of Good Hope. . . ." "Mr. Locke was precisely the writer to check so mischievous an epidemic of imaginary and spurious philosophy, by a well-timed satire, "out-heroding Herod" in its imaginative creations, supplying to satiety **the morbid appetite for scientific wonders then universally raging.**"

(continued)

Table 3.1 *(continued)*

Topos (Collocative Series)	Expectation	Examples
Plausibility (plausible, probable, shown by a writer on optics)	"If it seems like it could happen, it probably did."	"[T]he description of Sir John's recently improved instruments, the principle on which the inestimable improvements were founded, the account of the wonderful discoveries in the moon, &c., are all **probable and plausible**, and have an air of intense verisimilitude." "Educated in the strictest school of mathematical and inductive science, zealously devoted to its studies and pursuits, . . . Mr. Locke was precisely the writer to check so mischievous an epidemic of imaginary and spurious philosophy, by a well-timed satire, . . . maintaining a pretty firm grasp upon the credulity of nearly all but the thoroughly taught minds, by its **plausible** display of scientific erudition."
Verisimilitude (verisimilitude, vraisemblance, manner of its relation, well put together, written with consummate ability)	"Stories that sound like true science reports probably are."	The great effect wrought upon the public mind is referable, . . . to the exquisite **vraisemblance** of the narration." "[T]he principle on which the inestimable improvements were founded, the account of the wonderful discoveries in the moon, &c., are all probable and plausible, and have an air of intense **verisimilitude**."
Detail (scientific details, display of scientific erudition, account, description, particulars)	"Technical detail is a good indicator of truth."	"The writer (Dr. Andrew Grant) displays the most extensive and accurate knowledge of astronomy, and the **description** of Sir John's recently improved instruments, the principle on which the inestimable improvements were founded, the account of the wonderful discoveries in the moon, &c., are all probable and plausible, and have an air of intense verisimilitude." "In sober truth, if this account is true, it is most enormously wonderful, and if it is a fable, the manner of its relation, with all its **scientific details**, names of persons employed, and the beauty of its glowing descriptions, will give this ingenious history a place with 'Gulliver's Travels' and 'Robinson Crusoe.'"
Novelty (novelty, new era, discovery, announced, improvements, added a stock of knowledge)	"New discoveries are highly valued and probably true."	"The great effect wrought upon the public mind is referable, first, to the **novelty** of the idea." "The promulgation of these discoveries creates a **new era** in astronomy and science generally."

(continued)

Table 3.1 (*continued*)

Topos (Collocative Series)	Expectation	Examples
Authority (writer, Dr. Andrew Grant, learned astronomer, allow him to tell his story, Sir John)	"The author or authority's previous reputation holds."	"The **writer (Dr. Andrew Grant)** displays the most extensive and accurate knowledge of astronomy." "After all, however, our doubts and incredulity may be a wrong to the **learned astronomer**, and the circumstances of this wonderful discovery may be correct. Let us do him justice, and **allow him to tell his story** in his own way."
Medium (origin, journal, publish, republication, reprint, paper, the *Sun*, the *Edinburgh Journal of Science*, supplement)	"The previous reputation of the medium holds."	"As it professed to be a **republication** from the Edinburgh **Journal** of Science, it was some time before many persons, except professors of natural philosophy, thought of doubting its truth." "Our enterprising neighbors of the ***Sun***, we are pleased to learn, are likely to enjoy a rich reward from the late lunar discoveries. They deserve all they receive from the public—'they are worthy.'"
Internal coherence (intrinsic evidence, well put together, blunders, doubts and incredulity)	"If a story's claims are logically consistent, it is probably true."	"It appears to carry **intrusive evidence** of being an authentic document." "The singular **blunders** to which I have referred being properly understood, we shall have all the better reason for wonder at the prodigious success of the hoax."
Corroboration (nowhere but the penny papers, corroboration, without a word of notice)	"If a story is corroborated by several news sources, it is probably true."	"Without referring to the monstrosities of the story itself, can any one suppose for a moment that such preparations as are described, should have been made **without a word of notice** in the English papers?" "The only moral which we can derive from the whole matter is, that we must be more careful hereafter in believing everything which puts on the semblance of verity—especially if we find it **nowhere but in the penny papers.**"

These expectations governed the debate over the truth of the Moon Hoax. Because of the small number of archived responses, we will look first at popular science articles and their parodies to see if these expectations were met in the generic literature of the time. Then we will compare the expectations to Poe's hypotheses about which reader priorities and expectations drove the wide credence leant the Moon Hoax. Finally, we will look through all of these data sources for evidence of conflicts that produce rankings of the expectations in terms of their importance to readers. As a reminder, what we are aiming for, after we compare contemporary reactions to all of the hoaxes in the antebellum period, is a common reading filter, a ranked set of expectations that antebellum popular science newsreaders held in common.

Parody of the Moon Hoax

In the heat of the public debate over the hoax, three newspapers mounted a different sort of critique of Locke's story—parodies. The New York *Daily Advertiser* ran a letter from "Lock Loco" claiming that Herschel had plagiarized his lunar findings; the *Transcript* ran a story over two days that presented corroborating evidence for Herschel's findings from one "Captain Tarbox," and finally, on 2 September 1835, the *Herald* printed "A BET-

Table 3.2 Salient Features of "Moon Hoax" Parodied by *Herald*

Feature of "Moon Hoax"	Feature's "Double" in *Herald* Parody
Credentials/**authority** of **foreign** scientists	In the title: "BY HERSCHELL, THE GRANDSON, GRANDSON, L.L.D., F.R.S., R.F.L, P.Q.R., &C. &C. &C."
Locke's grandiose metaphors like "cloak of the Zodiack"	Phrases such as "stellar diadems" and "more numerous than the sparks which escape from a blacksmith's forge"
Verisimilar astronomical jargon	"hydro, philo, solar, high pressure steam telescope."
Locke's use of real-life analogies to make sense of immense astronomical distances and figures	"Latitude and longitude can be determined, in less time than an alderman could swallow a basin of turtle soup."
the "weight of scientific **detail**"	"Herschell then tasted the water of said ocean, by means of a very long hydrostatic tube, attached to the telescope. It has a very curious taste. He found it was composed of the following mixture, viz: 2 parts of lemonade, 1 part printer's ink, 1–2 parts mint julep, 1–2 parts flower of brimstone. There was also a slight tincture of blue vitriol."[56]

TER STORY.—MOST WONDERFUL AND ASTOUNDING DISCOVERIES, BY HERSCHELL, THE GRANDSON, L.L.D., F.R.S., R.F.L, P.Q.R., &C. &C. &C." This parody forms a good standard against which to compare our collected expectations. A parody, like a hoax, must foreground what is salient to the reader to achieve its effect. If a reader did not notice a particular feature in the original, the parody cannot achieve a comic effect by mimicking it. So the *Herald* parody of Locke's story is a good barometer of what readers paid attention to in the story. This parody mocks five particular features of Locke's tale, as diagrammed in table 3.2.

The *Herald*'s parody thus corroborates at least four of the popular science newsreading expectations that surfaced in reader responses to the hoax: authority, foreign prestige, verisimilitude ("sounding" scientific), and scientific detail. The parody also suggests that readers would immediately recognize as key features of contemporary popular science news bombastic metaphors and real-life analogies. To confirm these last two hypotheses, we need to look at a sample of articles from the time to see if these expectations are met and perhaps to discover other expectations that readers of Locke's hoax would have developed quasi-unconsciously through their reading of science news.

Science Writing in America, 1830–1845

Antebellum newsreaders knew what to expect from science writing not because they were educated to view it in a certain way, but because they read it and developed assumptions based on their repeated experiences with it. It is well documented that readers employ schemata (structured assumptions about speech and reading activities) to save cognitive effort.[57] The more assumptions that can be made about a text based on past experience with texts like it, the greater the processing capacity that can be dedicated to remembering new information in the text and evaluating it appropriately. Genre is probably the usual arena for exploring these assumptions, since a genre is itself a sort of schema or pattern of rhetorical moves that has been codified in medium, format, style, and content over the course of repeated interactions between author and reader through a certain textual medium. As discussed in the introduction, just this sort of feedback loop has operated in the development of the research article, according to Bazerman's and Berkenkotter and Huckin's work.

To try to discern the role of the contemporary media in this feedback loop, I examined a small sample of eleven antebellum newspapers and literary magazines for two purposes: first, simply to count how many and what kinds of science articles appeared in them, and second, to look for the bombastic metaphors and analogies parodied by the *Herald* along with any similar rhetorical patterns that might have developed into generic expectations through repeated reading.

Table 3.3 Antebellum Media Surveyed with Number of Science Articles per Issue

Magazine/Newspaper	Total Science Articles
Scientific American	80
Family Magazine	60
American Journal of Science	28
NY Sun	15
New York Herald	14
American (Whig) Review	7
Albany Argus	5
North American Review	2
New Yorker	1
Southern Literary Messenger	1
Broadway Journal	1

The eleven newspapers and magazines selected are listed in table 3.3. They were chosen either because Poe specifically mentioned having read science in them or because Frank Luther Mott listed them as examples of general science magazines in *A History of American Magazines: 1741–1850*.[58] For each newspaper or magazine, I selected a sample issue during the range of years of Poe's writing career, roughly 1830 to 1850. Whenever possible, an issue was selected from 1835, the year of the contest of Poe's and Locke's hoaxes. Within each issue selected, all of the articles and advertisements concerning issues of science and technology were categorized and counted to provide a snapshot of popular or general science reading in antebellum America. I found three major types of science writing:

- technology articles
- short, popular science blurbs and items
- longer articles treating "pure" science either as experimental report or educational piece

The definitional difference between science and technology is largely a product-driven one: if the article treated an invention, medicine, or machine, it was counted as technological. If it treated principles of science (including "pseudo sciences" such as mesmerism) or the history of the sciences, I counted it scientific. Within these three categories, several subcategories were discernable.

- metacommentary: articles specifically treating the state of science and/or technology in America

- blurb/factoid: short (one hundred words or fewer) items announcing a new scientific or technological discovery or simply stating a "gee whiz" fact about science
- ad
- joke
- poem
- sensation/spectacle announcement
- educational item
- how-to: treats technological or practical procedures for the lay reader
- home experiment: designed so the reader could observe a more abstract scientific principle at work

Within the category of longer "pure" science articles, three subtypes emerged:

- experimental reports
- observations
- reviews of science books.

The results of the survey are presented in tables 3.3 and 3.4.

Table 3.4 Distribution of Categories of Science Articles across Media Sample

Major Category	Subcategory	Totals for Subcategories	Subcategory % of Total Science Articles	Major Category % of Total Science Articles
Pop. Sci.	Meta	4	2.5%	
	Poem	5	3.1%	
	Spectacle	1	.6%	
	Joke	7	4.3%	
	Blurb	27	16.5%	27%
Pop Tech.	Educational	5	3.1%	
	Ad	47	28.8%	
	How-to	3	1.8%	
	Blurb	9	5.5%	
	Invention	11	6.7%	46%
"Pure" Sci.	Educational[59]	24.2	14.8%	
	Experiment	2	1.2%	
	Observation	10	6.1%	
	Review	8	4.9%	27%

These results show, unsurprisingly, that the more specialized science journals carried the most science pieces. The dominance of the popular technology category is due to the prevalence of ads, which accounted for almost 29 percent of all science articles, the largest "market share" of any of the subcategories. The strong showing of educational pieces (17.9 percent total) is slightly misleading as these pieces were almost entirely confined to three of the eleven journals in the sample—the *Scientific American*, the *American Journal of Science*, and the general education *Family Magazine*—and thus were not evenly distributed throughout the media, as were the ads and the blurb/factoids, which held a 16.5 percent market share. In these shorter pieces, science was marketed to the reader as a vendor of goods and services—technologies, in other words, or, as in the case of the blurb/factoids, as "gee whiz" entertainment. This sample from the media corroborates Robert Bruce's study of antebellum science, *The Launching of Modern American Science*. In a review of the letters and journals of American scientists of the era, Bruce found that they felt a constant pressure to produce technologies; many felt they had to be tied to industry in order to get even the most basic funding for their research.[60] The predominance of articles selling or advertising science here in part confirms the anxieties of these scientists.

An examination of the rhetoric of the longer articles bears out these observations to some extent, as science was figured as solving problems that faced average Americans, whether abstractly or concretely via technologies. While no two articles share exactly the same rhetoric, there are some striking recurring rhetorical features, summarized in tables 3.5 through 3.7 with examples.

Overall, the pieces seem to be structured along a problem/solution line, the problem in many cases being an "ignorance" of some scientific or technological principle, and the solution being a way to improve the average American lifestyle by application of that principle or technology. This finding is not novel; a long association between the problem/solution topos and science popularization has been documented by Charles Bazerman, John Swales, and Jeanne Fahnestock.[61] Fahnestock, in particular, motivates the adoption of this topos by popularizers as an epideictic turn—a desire to transform the basically forensic work of science into an immediate concern explicitly connected to the lay audience's values. A related way to motivate the use of the problem/solution topos is to view it as a narrative of control, where some untamable natural force becomes domesticated through American scientific and technological methods—certainly also an occasion for an *encomium* of ingenious native scientists.

Narrowing our focus from the global structuring principles of the texts in the sample, a few important local patterns can be discerned. The first lies in the rhetoric of the introductions or opening statements of these

Table 3.5 Typical Opening of Antebellum Science Articles

Rhetorical Feature	Examples (with features emphasized)
Mystery" opening • emphasizes **sensation** of phenomenon with words such as *wonder*, *enigma*, *mystery*, *puzzle*, *awe-inspiring*, *amazing*, etc. • often employs rhetorical questions to emphasize "mystery" aspect of phenomenon. • often implies that the phenomenon is a matter of almost universal attention and wonder.	"Every person of a reflecting mind must have often asked himself the question, *what are shooting stars*? The suddenness of their appearance, the rapidity of their motions, their *brilliancy*, the trains which they frequently leave behind them are well calculated to *awaken curiosity*." *AJS* "Few subjects of improvement have received *more attention* for the last twenty years than this, and it is with many people a matter of *astonishment*, that as late as within fifty years, and in most enlightened parts of this country, chimneys have been erected with fire places in which more than twelve times the fuel was required to be consumed in order to warm the room, that is now required for the same, or an equal purpose, in a modern approved stove." *Scientific American* "CURIOUS EXPERIMENT—Last Saturday a *novel* sight was seen in our harbor." *Herald* "This instrument [hydro-oxygen microscope] presents to our view a world of *wonders*. Its magnifying powers are *astonishingly* great." *Herald*

Table 3.6 Typical Structuring of "Problem" Phase of Antebellum Science Articles

Problem: ignorance on subject • typically follows the "mystery" opening immediately • claims that virtually no one knows the causes of the phenomenon • occasionally provides a folk explanation for phenomenon as comic relief	"and in *the absence of definite knowledge* respecting them [shooting stars], it is not perhaps strange that we have been favored with *an abundance of speculation and crude conjecture.*" *AJS* "*Still there are those who appear to understand little* of the true principles of economy in this respect." *Sci. Am.*
Transition: "smart people, however, know the answer." • immediately follows Ignorance segment • implies the existence of a solution/explanation for the phenomenon • often includes name-dropping of **cognoscenti**	"Among *the most extensive observations of this kind are those made by Professor Brandes of Lipsic*; and as they are but little known in this country, it may be acceptable to some readers of the Journal, to be furnished with an abstract of them." *AJS* "But we are glad to find many *who understand the thing better*, having looked into the theory of it." *Sci. Am.*

Table 3.7 Typical Structuring of "Solution" Phase of Antebellum Science Aarticles

Solution 1: **Details** of solution/explanation, often with illustration by example or scenario • illustrations often invite readers to **test out the solution/ explanation for themselves** • abstract principles are often made more familiar through the use of "real-life" **analogies** to things in the readers' experience.	"and the quantity of heat thus circulated, is in some measure proportionate to the velocity of this current. For *an illustrative experiment* on this subject, let any person select a spot on the surface of a stove that is red-hot, and blow with a common hand bellows directly on that spot for a few minutes." *Sci. Am.* "a large kite, at a great altitude, the line of which was fastened to the bows of the boat." *Herald*
Solution 2: use/benefit of principle • often introduced by a concluding sentential adverb such as *therefore*, *thus*, *then*, *so*, etc. • gives the pay-off of the solution/principle, in terms of money, efficiency, or occasionally, **new perspective or knowledge** • often contains sense of immediate benefit using words such as *soon*, *now*, *shortly*, *immediately*	"Now, *therefore*, we would recommend that in the construction of stoves, regard may be had to *facilitating a free circulation or current of air* over the exterior heated surface." *Sci. Am.* "We shall *soon* hear of our packet ships going over the Atlantic by the aid of kites at the rate of *a mile a minute*." *Herald*

science articles. Even in the more staid *American Journal of Science* articles, the opening rhetorical move is consistently mystery generating: "wonder," "astonishment," "curious," "novel," and "surprise" are repeated with uncanny consistency in all of the longer blurbs, educational pieces, and reports of discoveries.[62] This inflated introductory rhetoric mirrors the all-caps headlines of the science advertisements accompanying the articles and serves the same function—attracting and exciting attention.

The second local pattern visible in the sample is also familiar—the use of detail to convince the readers of the aptness of the author's solution to the acknowledged gap or lack. The *AJS* articles are the most detailed, likely due to their greater length and technicality. However, all of the popular science reports, including the blurbs/factoids, give details of functions, principles, and operations as part of their strategy to persuade the reader that their claims are true and valuable. This strategy is undoubtedly connected to those observed by Bazerman and Shapin in Enlightenment science articles, where manifold experimental detail and, often, meticulous engravings served to provide a virtual experience of true scientific witness. Indeed, even in these more popular scientific accounts, the reader could expect copious woodblock engravings that illustrated inventions, principles of astronomy or physics, or botanical or zoological specimens. Elizabeth Tebeaux argues in her study of Renaissance technical manuscripts that illustrations and text layouts, which helped the reader visualize topic structure, strengthened the perception that the text communicated the structure of reality.

A final local pattern obvious in the sample is the name-dropping of famous scientists—either as the authors of the novel work under discussion or in passing as colorful stamps of (usually Eurocentric) prestige. The ethos appeal of these citations is blatant, and in the case of foreign scientists, their distance from the context of publication made it harder for American readers to verify their work. This time-and-space buffer surely must have been attractive to hoaxers such as Locke and Poe.

All of these rhetorical moves listed above—the problem/solution/pay-off structure, the "mystery" opening, the dense detail, the engravings, the name-dropping of cognoscenti—since they appear fairly consistently throughout the different publications, could be considered expectations readers would hold, unconsciously or consciously, when approaching an article such as Locke's or Poe's hoax. The generic expectations mentioned in the reader responses of sensation, novelty, appeals to scientific authority, name-dropping foreign scientists, and copious scientific detail are clearly satisfied in the sample. Readers' expectations of internal coherence are matched by the relatively rigid problem-solution topos of the articles in the sample. The *Herald* parody's singling out of bombastic metaphor and analogy for mockery was obviously not accidental as we have seen many examples of these features in the "mystery" openings and the solutions phases of the typical antebellum popular science article.

Poe's and Locke's desires to hoax their readers surely grew in part out of their experience with the exaggerated contours of the popular science genre, both through reading it and, in Locke's case, through writing it. They put their intuitive understanding of the salient features of the genre to work when they wrote their hoaxes. In Poe's case, through his bitter analysis of the success of the Moon Hoax he made these intuitions explicit. His guesses about how his and Locke's readers read science news are instructive because both of them—Poe later than his colleague—refined their guesses into highly successful performances that fooled readers. His critique is our last stop before proceeding to define and rank reader expectations of antebellum science news.

Wrangling over the Moon Hoax

Poe attacked the Moon Hoax repeatedly in the years between its publication and his final burying of the hatchet in his 1846 portrait of Locke for his Literati of New York City series in *Godey's Lady's Book.* When he reprinted Hans Phaall as "The Unparalleled Adventure of One Hans Pfaall" in *Tales of the Grotesque and Arabesque* in 1840, he attached a "Note" to it comparing the two hoaxes and criticizing Locke's for its factual shortcomings. This essay was repeated in essence in one of Poe's freelance letters for the *Columbia Spy* in 1845 and again in the Literati portrait of Locke in 1846.

Poe's list of the errors in Locke's hoax is lengthy, and most of his complaints are easily corroborated, even now, by close examination of the text. Poe's list of errata includes these facts: that creatures seen on other planets would all appear upside down—if in fact you could see any more than the tops of their heads—and that Locke had made several simple multiplication errors in reporting the magnification power of Herschel's telescope and the relative sizes of the earth and moon. That Poe was able to catch all of these errors is a testament to his wide reading and acumen in astronomy and physics. In fact, both he and Locke benefited from the 1834 American reprint of Herschel's *Treatise on Astronomy* in preparing their hoaxes.[63] However, as Poe points out in his essay, Locke's scientific gaffes did not seem to have much effect on his readership, who suffered from the "gross ignorance which is so generally prevalent upon subjects of an astronomical nature".[64] Overall, Poe attributes the success of Locke's hoax to the following factors, as summarized in table 3.8.

Table 3.8 Poe's Characteristics of a Successful Hoax[65]

Content of Hoax	**Novelty:** being "first in the field" and thus finding readers unprepared to analyze the merits of the discovery because of lack of previous experience with it.
	Sensation: The "rich . . . fancy" of the amazing man bats, moon beavers, etc.
	Detail: Its "execution of details"; the minute observations about the construction and dimensions of the telescope and the labors of Herschel and his assistants to solve methodological problems.[66]
	Foreign and Corroboration: "Exclusive information from a foreign country"; by this Poe implies not only the strategy of making the hoax hard to confirm, but also the name-dropping of the famous foreign scientists and the illusion of corroboration by other journals.
	Plausibility: "Analogical truth" and "plausibility," which Poe believes were performed to perfection in his "Hans Phaall."
Presentation of Hoax in Its Medium	**Medium:** The reputation of its medium. The *Sun* was not yet known for printing hoaxes.
	Presentation: The "consummate tact with which the deception was brought forth"; Poe was undoubtedly recalling the blurb on page two of the 21 August 1835 *Sun* forecasting the discovery and the judicious periods of suspense between the parts of the report.
Style of Argumentation in Hoax	**Internal Coherence:** consistency of the argument, an issue separate from accuracy.
	Verisimilitude: "The exquisite vraisemblance of the narration"; more than one critic of Locke's hoax remarked on his elegant style, appropriate to such an awesome discovery, and the realistic sound of the research diaries "transcribed" by the *Edinburgh Journal of Science* and belonging to Herschel's chief assistant, Andrew Grant.

As we can see, most of these expectations match those already found in the other reader responses or via examination of genre texts. So, Poe's portrait of the "typical" reader resonates with the responses of actual readers. Still, Poe himself was not a typical popular science reader by virtue of his education and his bias as a hoaxer. His reconstructed reading expectations are only useful in that they have already been recovered from a wider sample of readers. However, Poe offers us something crucial that is difficult to get from a small sample of archived responses: ranking information about the reading expectations. His comments will help us begin to construct our common reading filter after we define our working set of expectations for antebellum popular science news so far.

Summary of Reader Expectations of Popular Science News

As discussed in chapter 2, a "reader expectation" is a sort of cognitive constraint on the process of creating meaning during the reading experience. In other words, the expectation functions as an assumption that favors the acceptance of certain interpretations of a text over others in constructing a "world view," which is the sum of all the beliefs the reader holds about the world.[67] That is why understanding hoaxing is crucial to understanding the dynamic between science and literature in nineteenth-century American social epistemology: hoaxing bears on the construction (reconstruction, from our standpoint) of the world in which readers of that era thought they lived. Those who believed Locke's hoax inhabited a different epistemic world than those who did not, a world where there were or could be man bats and moon bison. The words of Locke's story in the *Sun* were the only witness to the moon's surface, just as for many newsreaders today, words about Japan or the West or politics in Washington D.C. are their only contact with those parts of the world. It is this nascent power of witness to construct different realities for readers that Poe was beginning to exploit when he wrote Hans Phaall and when he spent so much time studying and worrying over the success of Locke's hoax.

All of the expectations that a reader has for a certain genre of text—popular science writing, in this case—interact to form a filter on all the possible meanings arising from the reading of any exemplar text from that genre. Only the meaning(s) that satisfies the greatest number of the reader's expectations will make it through the holes in the filter, will be allowed to change or "update" her world view. Walter Kintsch's work on the cognitive processes involved in comprehending text supports this kind of constraint-satisfaction model of reading. Kintsch's construction-integration (CI) model of reading presented in *Comprehension: A Paradigm of Cognition* (1998), is an extension of the comprehension model he developed with Teun van Dijk in the 1980s. The CI model maintains that all possible meanings of words and phrases are generated as a reader reads a sentence. Then, through a process of "spreading activation" only those meanings reinforced by the reader's prior reading experience and life experience are strengthened, where meanings irrelevant to the reader's goals and experiences are weakened and eventually die off.[68]

Kintsch's connectionist model of the reading process, while it provides support for a model of reading as an expectation-satisfaction process, mostly applies at a subconscious level, so it cannot cope with high-level, conscious decisions about truth-value of a text. Decisions about the truth of a text happen after the reader has already decided on the meanings of words and built a mental model of the text in her head; deciding about truth is deciding to what extent that model will interact with her model of the world. The constraint-satisfaction principle, however, can reasonably be expected to hold at high levels as well as at low levels in a connectionist system.

Below is a list of the expectations that were recuperated from the reader responses to the Moon Hoax and corroborated in Poe's analysis, parody, and a sample of popular science articles of the time. Notice that the expectations have the form of propositions; this is so that a possible interpretation arising from the text (say, "Herschel's telescope really works the way Locke says it does") can be tested for agreement with a particular proposition and either agree with it or violate it. These expectations, along with additional expectations recuperated through analysis of Twain's and De Quille's hoaxes, are also listed in the glossary of this book for reference:

Authority: The author or authority figure's previous reputation holds.

Corroboration: If a story is corroborated by several news sources, it is probably true.

Entertainment: Reading of popular science articles is for entertainment, not truth-value.

Foreign: Anything foreign is good and probably true.

Internal Coherence: If a story's claims are logically consistent, it is probably true.

Medium: The previous reputation of the medium holds.

Novelty: New discoveries are highly valued and probably true.

Plausibility: If it seems like it could happen, it probably did.

Popsci: Stories that sound like true science reports probably are. Subexpectations within this category are as follows:

Problem/Solution: Popular science reports are usually structured on a problem/solution topos. Related to internal coherence.

Long: Longer popular science articles are often given in installments.

Decoration: Popular science reports will often be decorated with bold headlines and woodcuts.

Mystery: Popular science reports often have a "mystery" opening signaled by words such as *wonders*

Ignorance/Wisdom: Popular science reports, after the opening, often employ an ignorance/wisdom antithesis to segue from problem to solution. Related to authority and foreign.

Detail: Popular science articles will often have a lot of technical detail.

Analogy: The details in a popular science article will often be explained with analogy to well-known phenomena.

Use: Popular science articles often finish with an evaluation of the benefit, physical or metaphysical, of the scientific principle/phenomenon.

Sensation: Sensational elements in a story have a high literary and truth-value.

Ranking Reading Expectations

Some reading expectations are clearly more powerful than others, because for all readers, certain assumptions override others. As a reminder from chapter 2, overridings or conflicts are usually signaled in reader responses by negative polarity items such as *however* or comparative terms such as *rather*. In this section we will establish as much of a common reading filter as we can by examining the loci of conflict in the reader responses to the Moon Hoax and synthesizing them wherever possible according to the method outlined in the previous chapter.

There are relatively few loci of conflict apparent in the primary reader responses. Many of the papers' editors, for example, just listed their highest-ranked reasons as justification for their decisions either to reprint or denounce it. The papers that reprinted it tended to cite Herschel's authority or the novelty or sensation of the piece. The papers that refused it tended to point to its problems with internal coherence or its lack of corroboration by European media. These comments give us an idea of what some of the editors' more important expectations were for science news, but without evidence of conflict, it is difficult to establish "winners" and "losers" in these interpretive games, and those kinds of victories and losses are crucial for establishing a ranked common reading filter.

Fortunately, some of the readers made their internal debates over the truth of the hoax explicit. For example, the New York *Sunday News* deferred its own judgment about the illogicality of Locke's report to Herschel's authority, whenever the astronomer could be reached for comment. That ranking of authority over plausibility and internal coherence was depicted in figures I.1 and tables I.1. Michael Floy Jr., a young New Yorker, recorded in his journal his own process of coming to believe Locke's story a hoax (topoi indexing expectations are bolded; conflict signals are underlined):

> A **great talk** concerning some **discoveries** in the moon by Sir John Herschell; not only trees and animals but even men have been **discovered** there. It is all a hoax, although the **story is well put together**. The author of these **wonders** says that an enormous lens of 30 feet diameter was constructed. He thought that would be a big enough lie in all conscience, but he should have said **a lens of 100 feet diameter, as it is shown by writers on optics** that such a diameter would be required to ascertain if any inhabitants in the Moon. Why not make a good lie at once? But it is utterly impossible to construct a lens of half that diameter, and therefore we may despair of ever ascertaining whether the moon be inhabited.[69]

Floy decides Locke's story is a hoax because his admiration of its internal coherence and verisimilitude to a popular science article is finally outweighed by the story's failure to provide a plausible telescope. His decision looks as follows (table 3.9) in OT. The fatal violation (!) of Floy's high-ranked expectation that true science stories are always plausible is what makes it impossible for him to believe the story. That is how we know plausibility is more important to him than internal coherence and popular science verisimilitude—because he negates his approval of the story's narrative appeal and tightness in order keep his expectation of plausibility in science stories inviolate. So far, we have two reader decisions that defeat internal coherence in favor of stronger expectations: plausibility in one case and authority in the other.

Table 3.9 Reader Michael Floy's Decision to Disbelieve Locke's Moon Story

	Plausibility	Internal Coherence	Popsci
TRUE	*!		
✓FALSE		*	*

Several of the editors mention the lack of corroboration by other news sources as their reason for denouncing Locke's story as a hoax. One of them, the editor of the New York *Commercial Advertiser*, gives us a glimpse into the debate that he experienced in deciding how to reprint the story.

> [W]e have copied on our first page the **marvelous** account which has been made the occasion of so much discourse lately, of certain **stupendous discoveries** in the moon, alleged to have been made by Sir John Herschel at the Cape of Good Hope. It is **well done**, and makes a pleasant piece of reading enough, especially for such as have a sufficient stock of credulity; but we can hardly understand how any man of common sense should read it without at once perceiving the deception. Without referring to the **monstrosities** of the story itself, can any one suppose for a moment that such preparations as are described, should have been made **without a word of notice in the English papers**?

This editor deemed the story's lack of corroboration too damning to be outweighed by the appeals of its sensation, novelty, and its strong resemblance to a true popsci. article. A graph of his decision appears in table 3.10. What this editor's responses add toward our project of constructing a common reading filter is that now we have seen two interpretive games in which popsci. has been defeated. It remains to be seen if this trend will hold in other reader responses.

While ferreting out most readers' rankings takes some work, Poe did our work for us by ranking his criteria for a successful hoax. Comments in all versions of his essay about the competition of his hoax with Locke's hoax indicate that Poe ranked four of the criteria (the novelty, sensation, presentation, and verisimilitude criteria from table 3.8) in order of the strength of their impact on readers:

> The singular **blunders** to which I have referred being properly understood, we shall have all the better reason for wonder at the prodigious success of the hoax. Not one person in ten discredited it, and (strangest point of all!) the doubters were chiefly

Table 3.10 NY *Commercial Advertiser* Editor's Decision to Disbelieve Locke's Moon Story

	Corroboration	Sensation	Novelty	Popsci.
TRUE	*!			
✓FALSE		**	*	*

> those who doubted without being able to say why—the ignorant, those uninformed in astronomy, people who would not believe because the thing was so novel, so entirely "out of the usual way." A grave professor of mathematics in a Virginian college told me seriously that he had no doubt of the truth of the whole affair! The great effect wrought upon the public mind is referable, **first**, to the **novelty of the idea; secondly**, to the **fancy-exciting and reason-repressing character** of the alleged discoveries; **thirdly**, to the consummate **tact** with which the deception was brought forth; **fourthly**, to the **exquisite vraisemblance** of the narration [emphasis mine].[70]

Poe feels overall that when it came to weighing the scientific inaccuracies of Locke's story against the novelty of moon bison and man bats, people allowed themselves to be swayed by the novel and sensational. While Poe separates out the way the story was presented from its resemblance otherwise to a true popular science article, none of the other contemporary readers make this distinction; so, from here on, presentation will be subsumed under popsci. The reader in table 3.11 ranks his/her expectations as Poe projects.

In the table the nine factual errors in Locke's story are counted with nine asterisks representing eight violations of plausibility and one of internal coherence. This tally, of course, assumes that the average reader recognized all nine errors, which was indeed the assumption Poe was making in the passage from his Literati essay just above. In spite of all the faulty evidence, Poe's projected reader believes Locke's story to be true simply because to consider it false would force the admission that something novel, sensational, and "verisimilar" is not true; in this reader's world view, the correlation between spectacular first impressions and truth cannot bear violation. The exclamation point on the chart indicates that all violations at this highest level of expectation are unacceptable (the convention is to mark unacceptability on the very first violation that renders the candidate interpretation unacceptable, and this is usually the weakest or right-most violation on a given level, as expectation strength increases from right to left).

Table 3.11 Graph of Decision by Poe's Projected Readers to Believe in Locke's Moon Bison

	Novelty	Sensation	Popsci.	Plausibility	Internal Coh.
✓ TRUE				********	*
FALSE	*	*	*!		

Thus, the candidate with more total violations actually wins in this case because of the very low value assigned to scientific accuracy by the reader. This accurately represents Poe's complaint about Locke's readers' values. Now a table of Poe's personal reading of the story would be almost exactly the reverse of the one above, with his precious "analogical truth" in the form of plausibility and consistency ranked very firmly over novelty, sensation, and "presentation/verisimilitude" (popsci.).

The number of interpretive "games" or decisions that could be played with these science newsreading expectations is potentially infinite. Some of the games we could play with Locke's hoax would bring other pairs of expectations into competition with each other. Entertainment is a particularly interesting constraint, because when it is in operation as an expectation—as evidenced by lack of negative polarity items in the text of the response and the substitution of sentential adverbs expressing conditionality, such as *if*, and *whether or not*—it forecloses on games about truth-value. Entertainment was the highest-ranked expectation for the two responders who read Locke's story for pleasure, enjoying its colorful images and turns of language, but who did not care if the details of the story were real or not. In terms of current theories of the psychological bases of fiction reading, these entertainment readers "decouple" the meanings arising from their reading of Locke's story from their world view.[71] Their worlds, like the worlds of disbelieving readers, do not alter to contain moon bison. What is different about the experiences of those two groups of readers, however, is that entertainment readers' realities were never "on the table," so to speak, because the entertainment expectation actually deactivates truth-value expectations such as authority and plausibility. Entertainment readers are reading fiction, not news, even if the story appears in a newspaper! This intriguing situation will be fleshed out further when we examine the *letturaturizzazione* of Poe's hoaxes.

Another worthwhile observation about these reading expectations is that they were employed both by the writers and by the readers of the hoaxes, because writing hoaxes is itself a game in reading readers and guessing their expectations. We have already seen Poe's guesses about his readers' ranking of their expectations. Another writerly example comes from the editors of the *Sun*. They introduced the first substantial installment of the hoax on 25 August with this note: "We are necessarily compelled to omit the more abstruse and mathematical parts of the extracts however important they may be as a demonstration of those which we have marked for publication; but even the latter cannot fail to excite more ardent curiosity and afford more sublime gratification than could be created and supplied by any thing short of a direct revelation from heaven." The editors knew scientific detail was crucial for floating the Moon Hoax, but they did not wish to bog down the sensational elements of the story in formulae, so they "omit[ted]" the "more abstruse and mathematical parts" for the general reader. This shows that in their writing strategy (which is also a game in reading their readers), sensation was ranked over detail. These writerly guesses are useful because they were successful to the degree that the writers' hoaxes were successful with readers. However, since they are secondary reactions (guesses about reactions), they must be counted less than primary reader reactions in constructing a common reading filter, especially when the secondary judgments contradict primary judgments. We will resolve just such a contradiction at the end of this chapter.

What we have collected now is a set of expectations—some of which interact and compete with each other, and some of which foreclose on others—that we can use as a basis for examining the rest of Poe's hoaxing attempts and readers' reactions to them. These expectations illuminate which of the readers' values were at stake in the writing and interpretation of the hoaxes, and they help explain the success or failure of the hoaxes with those readers. Once we establish these points of analysis, then we can finally turn to our major goals in this chapter: synthesizing a common reading filter for antebellum science newsreaders and determining what kind of game Poe was playing with his readership in his hoaxes, and why.

The Balloon-Hoax

Poe's next attempt at a hoax was on 13 April 1844, when an "Extra" to the regular Saturday edition of the New York *Sun* trumpeted the following in "magnificent capitals":

> [Astounding News by Express, via Norfolk!—The Atlantic Crossed in Three Days! Signal Triumph of Mr. Monck Mason's Flying Machine!—Arrival at Sullivan's Island, near Charleston, S.C., of Mr. Mason, Mr. Robert Holland, Mr. Henson, Mr. Harrison Ainsworth, and four others, in the Steering Balloon, "Victoria," after a Passage of Seventy-five Hours from Land to Land! Full Particulars of the Voyage!]

The original article continues, "The great problem is at length solved! The air, as well as the earth and the ocean, has been subdued by science, and will become a common and convenient highway for mankind."

Certainly, this opening must have satisfied readers' demands for novelty and sensation, as aeronautics was a hot topic in the popular scientific and technical articles of the late 1830s and early 1840s. British balloonist Monck Mason had recently published a memoir of his crossing of the English Channel in a balloon, and there was high anticipation that Mason would soon attempt something longer.

However, what is more significant is that Poe had worked harder on the rhetoric of his piece. Almost immediately apparent are several ways in which Poe has crafted the language of his new hoax to better match reader expectations: First, the tone of the opening, with its exclamation marks and use of adjectives such as "astounding," is completely in keeping with the mystery introductions of other science articles. The "bantering" tone of Hans Phaall is gone, replaced by a serious, if sensational, journalistic style. Also, notice that Poe immediately invokes the problem/solution paradigm current in the popular science reporting of the day with the sentence "The great problem is at length solved!"

The article goes on to imitate the ignorance phase of the problem/solution topos by detailing past failures in aeronautical experiments. To satisfy reader expectations about wisdom, Poe then name-drops the famed appellations of balloonist Monck Mason, whose diaries Poe copies for large sections of his report,[72] and of popular British historical romance writer Harrison Ainsworth. After that, the mechanical details of the balloon's construction are focused on minute extremes, and a decorative woodcut of the wedge-spiral propeller of the balloon is included. The account ends true to the reader's use expectations with an overblown assessment of the importance of the voyage and the speculation, "What magnificent events may ensue, it would be useless now to think of determining."[73]

Did readers respond better to this hoax that seemingly conformed better to their expectations of how a true science report should read? The short answer is yes, but complicating it is the fact that the only extant eyewitness to the reaction to the hoax is Poe himself, who undoubtedly exaggerates it in his 21 May 1844 letter for the *Columbia Spy*: "The 'Balloon-Hoax' made a far more intense sensation than anything of that character since the 'Moon-Story' of Locke. On the morning (Saturday) of its announcement, the whole square surrounding the 'Sun' building was literally besieged, blocked up—ingress and egress being alike impossible, from a period soon after sunrise until about two o'clock P.M."[74] The *Sun* did indeed sell a record number of copies of the Saturday extra containing the hoax, fifty-thousand according to the estimate in the Philadelphia *Saturday Courier*,[75] so Poe's jubilation is perhaps founded in reality. He admits that "of course there was great discrepancy of opinion as regards the authenticity of the story; but I observed that the more intelligent believed, while the rabble, for the most part, rejected the whole with disdain."[76] This is the same argument, curiously, that he made about Locke's hoax, but in this case he is using it to aggrandize his intelligence and the craftsmanship he invested in the scientific detail of this hoax. "As for internal evidence of falsehood, there is, positively, none—while the more generally accredited fable of Locke would not bear even momentary examination by the scientific. There is nothing put forth in the Balloon-Story which is not in keeping with the known facts of aeronautic experience—which might not really have occurred."[77] Poe claims further that he listened in on people's discussions of the extra, and the only quibbles he heard with the story were on account of the reputation of the *Sun*, since it had published the Moon Hoax, and the difficulty of having gotten news from Charleston so quickly. Indeed, the editors of the New York *American* record this criticism of the story's plausibility: "The express, which has hardly outstripped the ordinary mail, must also have brought along a woodcut of the balloon, as the *Sun* has the picture as well as the story—one as good as the other."[78] These doubts on the part of readers correspond to the medium and plausibility expectations and, depending on their strength with particular readers, might well have had the effect Poe reports. Table 3.12 presents the "filter" of expectations of readers such as the editors of the New York *American* who disbelieved the Balloon-Hoax on the basis of medium and plausibility. For these readers, the fatal violations were of medium and plausibility, equally ranked over popsci., or the story's "verisimilitude" in conforming to reader expectations of a popular science article. Even though denying the truth of the article meant negating the article's convincingly constructed scientific rhetoric, the *Sun*'s tarnished reputation and readers' background knowledge of the difficulty of getting mail so quickly from Charleston sway readers' decisions.

The Balloon-Hoax did not have the life of the Moon Hoax, however, and that is because Poe revealed himself as the perpetrator that afternoon. An acquaintance of Poe, New York journalist and "free love" advocate Thomas Low Nichols, wrote in his autobiography that Poe got drunk on wine and stood on the steps of the *Sun* building the very afternoon of the hoax's debut, shouting out to potential buyers of the extra that it was a fake.[79] A corroborating bit of evidence is that the *Sun* printed a retraction for the hoax two days after its publication, an admission that it had stubbornly refused to stoop to in the case of the Moon Hoax.[80] By this account, Poe witnessed the run-away success of his hoax and could not bear for its readers not to know that he was its creator. At most that admission cost him a day or two of entertainment, because the mails from Charleston almost certainly would have arrived on Monday or Tuesday, shutting the

Table 3.12 Decision about the Truth-Value of Poe's Balloon-Hoax by Readers Valuing Plausibility and the Reputation of the News Medium

	Medium	Plausibility	Popsci.
TRUE	*	*!	
✓FALSE			*

hoax down. Locke's hoax had benefited from a much longer window of play before Herschel could be reached for comment.

A couple of dynamics within the Balloon-Hoax event speak directly to two of the three primary research questions of this book. In partial response to the question, Why did literary authors choose to write science hoaxes? the Balloon-Hoax reveals that Poe used it as a stage on which to construct himself as a notorious expert; second, addressing the question, What is a hoax? we see that as Poe constructed his phantom balloon, he also developed a curious textual mechanics that would characterize not only his detective and science fiction, but also future scientific media hoaxes by other authors.

Professionalism, Expertise, and Hoaxing

The image of Poe gloating triumphantly over newsreaders gawking up at him with his hoax still open in their hands is a pose worth investigating. For one of the many rhetorical functions of a hoax is the public notoriety it constructs for its creator. Poe's willingness to step into that role on the front steps of the *Sun* suggests that it may have been a powerfully motivating factor in encouraging him to continue attempts at hoaxing. Scrupulously protecting his identity as a hoaxer, as did Locke, was not for Poe. Poe wished to be publicly recognized as a writer whose scientific expertise enabled him to beat scientists at their own game and to construct "discoveries" for the public that they could not tell from the real thing. Because of this motivating force, Poe's hoaxing is bound up with the history of professionalism and the construction of public expertise in America.

As mentioned above, Poe grumbled about the difficulty of making a living as a professional author and editor while scientists prospered. Although Poe chalks up this inequity to deficiencies in public taste—and so uses hoaxes to embarrass his readers for these deficiencies—the truth is that a whole complex of historical and economic factors had made it possible for both Poe and the scientists he resented to make their livings as professionals. William Charvat, in his study of the profession of authorship in America, says the most common definition of professionalism is getting paid to do something. As a corrective to this truism, he notes that even before James Fenimore Cooper became the first financially successful author in American history, writers such as Susanna Rowson and Joel Barlow could be counted professionals even in the absence of economic success because they publicly claimed the vocation of an author.[81] Their ethos as professional writers, according to Charvat, qualified them as much as or more than did their bank balances.

The expansion of the publishing industry in the 1820s finally created an economic niche to match the rhetorical claim staking of professional authors such as Cooper. At the same time, as discussed in chapter 1, rapid industrialization created similar economic niches for scientists and engineers as professionals. While the term *professional* might constitute their economic identity, the term *expert* described their social and epistemological identity. These experts embodied the ways in which abstruse fields of study such as electromagnetics, physics, and chemistry were becoming assimilated into the human social system; they personified the taming of nature. As such, they served as oracles through which the lay public could interrogate the natural world.

But this culture of expertise did not develop without resistance. The mistrust of elitism that characterized Jacksonian politics derived in part from a deep public discomfort with secondhand access to crucial knowledge and a suspicion of private agendas on the part of the professionals who dispensed it. Theodore Porter, in *Trust in Numbers*, marks these factors as signal for the development of a culture of objectivity. Porter defines objectivity as "technologies of trust" in place of personal trust. Measurements, standards, and rules, also enabled by an industrial urbanized economy, allowed lay people to determine the extent of their assets and experiences for themselves without the questionable intervention of experts. In these ways, a culture of objectivity grew up and stood opposed to the culture of expertise.[82]

Poe's "Balloon-Hoax" fed directly off these tensions between objectivity and expertise in his culture and allowed him to construct himself as a notorious (read, countercultural) public expert. By publicly revealing his hoax he demonstrated that the objective "rules" his readers had used to determine the truth of his science report for themselves had let them down. His public humiliation of them formed an implicit argument that they could not trust themselves to understand science, and they could not even trust scientists, since Poe had clearly bested them at their own game. Poe and his imaginative artistic colleagues alone were qualified as experts in social truth. If readers wanted the truth, they would have to go through professional artists like him. By this chain of implications, Poe's hoax struck against Baconian empiricism and objectivity in favor of the culture of expertise, with himself at the center of it. In the next chapters, we will see this function of hoaxing reasserting itself as a motivation for the hoaxes of Mark Twain, Dan De Quille, and Alan Sokal.

Textual Mechanics

Another striking element of the Balloon-Hoax, and of Hans Phaall as well, is Poe's almost obsessive attention to mechanics—mechanics of the machines in the story, mechanics of the story itself. In both hoaxes mechanism fairly overwhelms the plot; one could argue the details about the balloons *are* the plot. With the Balloon-Hoax, however, the very *rhetoric* of Poe's hoaxing is becoming mechanical. In 1836 he had laboriously detailed and exposed Maelzel's chess-playing automaton as a fake; the gears and cogs in the cabinet were merely a front for a chess expert who, from his cramped position, manipulated the automaton's arm via levers to move the pieces. Poe was actually practicing his own brand of legerdemain in this exposé, as he lifted most of it directly from magician David Brewster.[83]

Writing hoaxes seems to be an extension of this engineering rhetoric. With each new attempt (including his exposé of Locke's hoax), Poe tinkers. He tests the rhetorical impact of these "*jeux d'esprit*" on the reader and then tweaks an appeal here, adjusts the tone there. He even begins to construct his hoaxes by salvage, recycling, in the case of the Balloon-Hoax, whole parts of Monck Mason's *Account of a Late Aeronautical Expedition from London to Weilberg* as well as a contemporary science report "Remarks on the Ellipsoidal Balloon propelled by the Archimedean Screw, described as the New Aerial Machine."[84] Poe's prose in these hoaxes builds flying machines. His hoaxes are flying machines, after a fashion. Contemporary William Griggs appraised Locke's hoax as a mere "balloon" that was mistaken for a while for a "celestial luminary."[85] Poe's hoaxes fit this analogy to the letter. In constructing the *Victoria*, Poe makes his readers move with him, a step at a time, through the building and piloting of the balloon and thus makes them coengineers in his hoaxes. Here is the painstaking detail with which Poe describes the principle means of locomotion for the *Victoria*, the Archimedean screw:

> The *screw* consists of an *axis* of hollow brass tube, eighteen inches in length, through which, upon a semi-spiral inclined at fifteen degrees, pass a series of steel wire *radii*, two feet long, and thus projecting a foot on either side. These *radii* are connected at the outer extremities by two bands of flattened wire—the *whole* in this manner forming the framework of the *screw*, which is completed by a covering of oiled silk cut into gores, and tightened so as to present a tolerably uniform surface. At each end of its *axis* this *screw* is supported by pillars of hollow brass *tube* descending from the hoop. In the lower ends of these *tubes* are holes in which the pivots of the *axis* revolve. From the end of the *axis* which is next the *car*, proceeds a shaft of steel, connecting the *screw* with the pinion of a piece of *spring* machinery fixed in the *car*. By the operation of this *spring*, the *screw* is made to revolve with great rapidity, communicating a progressive motion to the *whole*. [emphasis mine.][86]

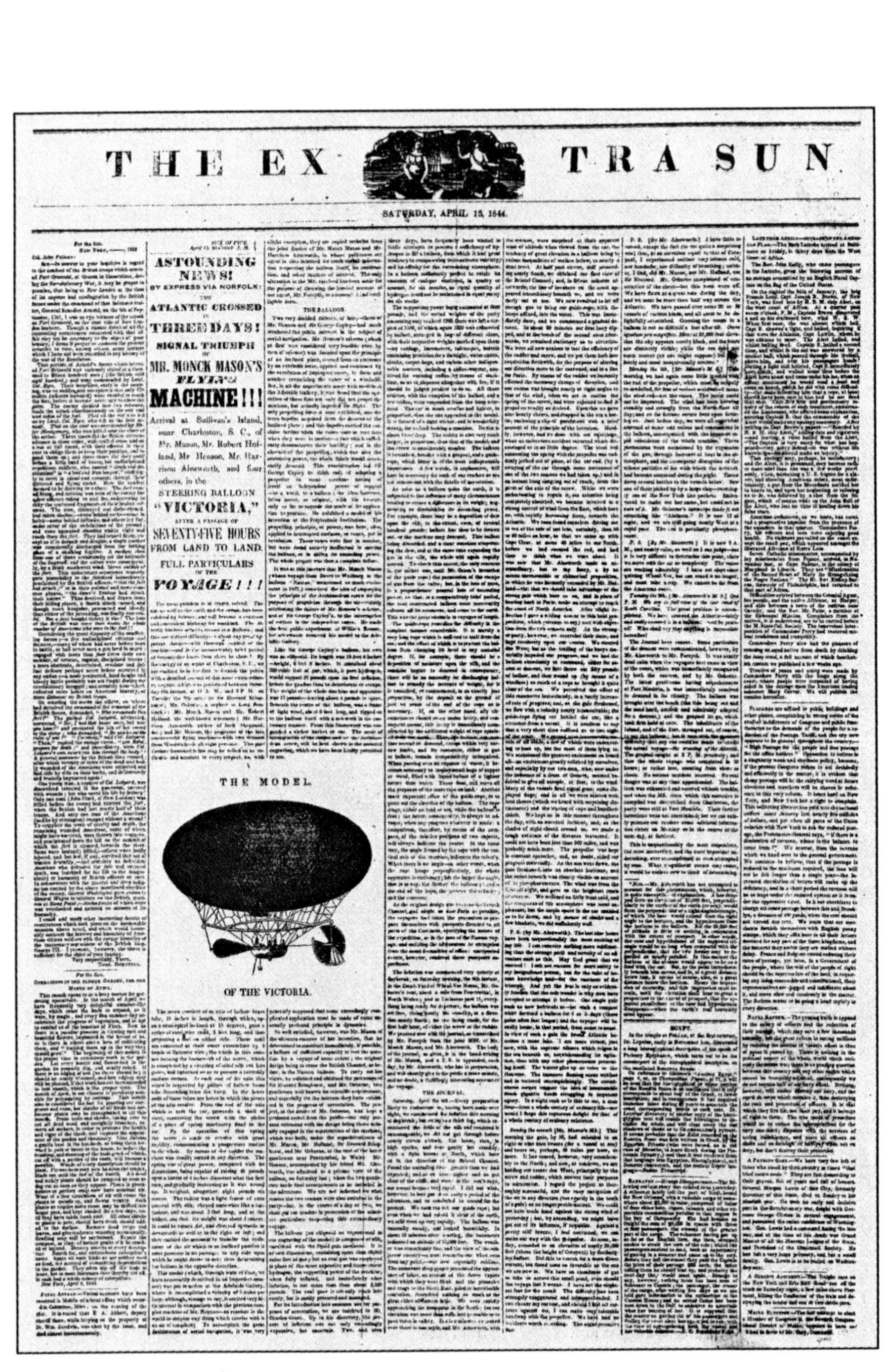

THE EXTRA SUN.

SATURDAY, APRIL 13, 1844.

ASTOUNDING NEWS!

BY EXPRESS VIA NORFOLK:

THE ATLANTIC CROSSED IN THREE DAYS!

SIGNAL TRIUMPH OF MR. MONCK MASON'S FLYING MACHINE!!!

Arrival at Sullivan's Island, near Charleston, S. C., of Mr. Mason, Mr. Robert Holland, Mr. Henson, Mr. Harrison Ainsworth, and four others, in the STEERING BALLOON "VICTORIA," AFTER A PASSAGE OF SEVENTY-FIVE HOURS FROM LAND TO LAND.

FULL PARTICULARS OF THE VOYAGE!!!

THE BALLOON.

THE MODEL OF THE VICTORIA.

THE JOURNAL.

EGYPT.

Figure 3.3 The First Page of Poe's Balloon Hoax. *Courtesy, American Antiquarian Society*

This lengthy schematic is approximately one-eighth of the total amount of technical description in the story. I have italicized certain words in the passage that are repeated, and by focusing on them, we can appreciate some of Poe's rhetorical mechanics. First, the screw is constructed as a unit. Poe uses "whole" to alert the reader that what they have read to that point completes the construction of the screw proper. The second mention of the "whole" at the end of the passage refers to the balloon and tells the reader how the screw fits and functions as a part of the balloon. In finer detail, the screw is constructed rhetorically a step at a time, with each component fitting into the next until the function of the entire screw in context is apparent. "Screw," "axis," "radii," "tube," "spring," and "car" are repeated, and their repetitions are layered in with each other such that each piece of the screw leads the reader to the next piece in the assembly. The screw is literally constructed and put into play before the reader's eyes. Poe's readers are reading an engineering schematic. Recall, too, that this story was accompanied by an engraving of the Archimedean screw, so at each stage in the construction the reader could compare his mental model of what he had constructed so far to the graphic.

Poe's casting of the reader as coengineer is not insignificant. Not only does the reader mentally construct a balloon as she reads, but she also constructs the hoax, because the balloon—its very existence, its successful functioning—is the hoax. Notice Poe's use of the verb "communicate" to describe the energy transfer in the *Victoria*'s propeller. Both the balloon and the text are performing the same function—implicating the reader in pulling the wool over her own eyes, as by the time she is done reading, she can see nothing but the gears, shafts, and pinions spinning in the text, cleverly hiding, while also mechanically accomplishing, Poe's real agenda. Poe is practicing a mechanics of rhetoric with these hoaxes that will in *Eureka* influence his very conception of how reality works.

"The Facts in the Case of M. Valdemar"

In the December 1845 edition of *The American Review: A Whig Journal*, subscribers read "The Facts in the Case of M. Valdemar" by Edgar A. Poe, which began as follows:

> Of course I shall not pretend to consider it any matter for wonder, that the extraordinary case of M. Valdemar has excited discussion. It would have been a miracle had it not—especially under the circumstances. Through the desire of all parties concerned to keep the affair from the public, at least for the present, or until we had farther opportunities for investigation—through our endeavors to effect this—a garbled or exaggerated account made its way into society, and became the source of many unpleasant misrepresentations, and, very naturally, of a great deal of disbelief.

> It is now rendered necessary that I give the facts—as far as I comprehend them myself. They are, succinctly, these:[87]

Roland Barthes has done a close reading of the language Poe uses in this piece, and in the opening, Barthes finds two "codes" operating (codes are complex constraints on interpretations similar to the "filters" of reading expectations we have been examining): the code of the enigma, and the code of science.[88]

The enigma code, as Barthes describes it, is very similar to the requirements for the mystery expectation. An enigma operates by lack—astonishment, wonder, ignorance. Barthes points to the linguistic cue of the definite article "the" introducing "the extraordinary case" and "the affair"; the definite article registers a linguistic presupposition that the case/affair exists in the world of Poe's readers and that they should be aware of it. "The facts" and "the circumstances" presuppose an enigma or misunderstanding that will now be cleared up.

The noun phrase "the facts" also invokes the code of science, through which, according to Barthes, scientists try to position their endeavors outside the realm of literature. The code of science is a concentrated attempt by scientists writing science to supplant symbols—and, indeed, all symbolic construction of meaning where one thing stands for another—with "facts" which simply are and communicate truth immediately. Of course, Barthes would have probably been the first to agree that science has its own symbols, but his point here is that Poe uses language in the opening of M. Valdemar the way that scientists use language in order to convince his readers to put their trust in his eyewitness account of the "extraordinary case." The enigma code and the scientific code are in competition with each other throughout this story, inasmuch as enigmas disguise truth behind symbols or clues, while science seeks to channel truth asymbolically through facts, according to Barthes.

Poe's story continues past this enigmatic opening to detail a case of mesmeric suspended animation, in which the dying M. Valdemar is hypnotized at the point of death. When he is taken out of the trance state, months later, he immediately decomposes in his bed, having been in fact dead the entire time. Poe discusses Valdemar's medical condition prior to his death in great scientific detail and lists the names of several important doctors attending the case but excises all but the first letters of their names: Dr. D———, Dr. L———, etc. This device is an interesting variant of the name-dropping that is characteristic of the wisdom, authority, and foreign expectations from popular science writings. This elision possibly derives from eighteenth- and nineteenth-century conventions of journaling, dramatized in novels of social exposé such as Laurence Sterne's *Sentimental Journey*. The design of Poe's use of the convention was probably to convince readers that this case was so sensational and extraordinary that the famous people involved (doctors, in this case) wished their involvement in it to be kept private in order to insulate their public reputations. Poe refers to

Valdemar himself as "well-known" as a translator of various famous European works and provides a list of phony credentials.

Poe obviously worked to satisfy the popsci. cluster of expectations of "verisimilitude" because his style matches very closely the medical case studies at this time, particularly mesmeric case studies that include dialogue with the mesmerized subject. It is likely that Poe consulted several sources in giving Valdemar an authentic rhetorical feel, particularly a reprint of the Rev. Gibson Smith's pamphlet "Lectures on Clairmativeness, of Human Magnetism" featuring the work of clairvoyant Andrew Jackson Davis.[89]

The story was apparently believed widely and reprinted copiously in England, even in the London *Times*.[90] The Edgar Allan Poe Society has published online two letters from readers that contain references to the hoax. The first is from Elizabeth B. Barrett; here is the apt section (again, I have bolded phrases connected to reader expectations of science news and have underlined elements indexing conflict):

> Then there is a tale of yours ("The Case of M. Valdemar") which I do not find in this volume, but which is going the round of the newspapers, about mesmerism, **throwing us all into a "most admired disorder,"** and dreadful **doubts** as to whether "it can be true," as the children say of **ghost stories**. The certain thing in the tale in question is the power of the writer, and the faculty he has of **making horrible improbabilities seem near and familiar.**[91]

This is one of the first cases so far in which we have seen a reader struggling with a truth decision about a hoax and not coming to a conclusion. In these cases, the expectations in play are held in suspension. Barrett mentions the sensation of the piece "admired disorder" and the verisimilitude of its writing "making horrible improbabilities seem near and familiar"; against that, causing "doubts" is the implausible topic of the story, making it fit for a "ghost story." The violations all balance, and without ranking information, a decision cannot be cast.

A similar dilemma emerges from the response of Arch Ramsay from Stonehaven, Scotland:

Table 3.13 Suspended Decision about the Truth-Value of Poe's Facts in the Case of M. Valdemar by E. B. Barrett

	Sensation	Plausibility	Popsci.
TRUE		**	
FALSE	*		*

> As a believer in Mesmerism I respectfully take the liberty of addressing you to know, if a pamphlet lately published in London (by Short & Co., Bloomsbury) under the **authority** of your name & entitled Mesmerim, in Articulo-Mortis, is genuine.
>
> It **details** an acc't of some most **extraordinary** circumstances, connected with the death of a Mr M Valdemar under mesmeric influence, by you. Hoax has been **emphatically pronounced upon the pamphlet by all who have seen it here,** & for the sake of the Science & of truth **a note from you on the subject** would truly oblige.[92]

Ramsay's response is particularly interesting. He cannot decide the truth of the hoax based on the information he has, so he appeals to Poe to satisfy stronger constraints on his decision process—corroboration and authority. Ramsay would like to believe the story. His expectations of scientific detail and sensation are satisfied, but his stronger-ranked belief that stories like this must be corroborated is currently violated by the general consensus from the English media is that the story is a hoax. He has not given up, however, because of his equally strong belief that mesmerism is plausible. He is hoping to receive authoritative word from Poe to counter the negative corroboration he has received and make it possible for him to publicly defend his belief in the truth of M. Valdemar.

Poe's response to Ramsay was strangely cagey in contrast to his gloating revelation of the Balloon-Hoax:

> "Hoax" is precisely the word suited to M. Valdemar's case. The story appeared originally in "The American Review," a Monthly Magazine, published in this city. The London papers, commencing with the "Morning Post" and the "Popular Record of Science," took up the theme. The article was generally copied in England and is now circulating in France. Some few persons believe it—but I do not—and don't you.[93]

Poe curiously adopts the stance not of an author but of a bystander lacking the authority to pronounce definitively on the truth of the hoax when he writes, "Some few persons believe it—but I do not—and don't you."

Table 3.14 Suspended Decision about the Truth-Value of Poe's Facts in the Case of M. Valdemar by A. Ramsay

	Corroboration	Authority	Plausibility	Detail	Sensation
(✓)TRUE	*				
FALSE	(*)	(*)	*	*	*

Whether Poe actually intended M. Valdemar as a hoax when he wrote it is unclear. He claims he did not, in a letter to Evert Duykinck in 1848. However, it is quite possible Poe was continuing his deliberate tinkering with the expectations of his reading public with this tale. It certainly seems that way, given that he reperformed all of the popsci. reader expectations performed in the Balloon-Hoax. In addition, Poe satisfies the sensation and novelty expectations in spades because the effect of Valdemar speaking from the dead and rotting away in front of the reader was very unusual for a staid political journal such as the *Whig Review*. Also, as Barrett testifies, Poe chooses for this tale another hot topic on the level of balloon aviation. Roland Barthes points out in one of the rare "crossings" of history into his analysis that 1845 was perhaps the height of the madness and "scientific illusion" concerning mesmerism (called "magnetism" in Europe).[94] This is one of the many ways in which we are hampered by our historical position when we read Poe's hoaxes. While to us hypnotism and mesmerism are the last things we would call plausible science, the situation was markedly different in Poe's time, when painless births and surgeries were allegedly being performed under hypnosis.[95]

In spite of all this evidence for deliberate design, it is just as likely, however, that in M. Valdemar Poe may have been experimenting with mesmerism in fiction along the vein of his detective stories, since he demystifies the enigma of M. Valdemar's case step by step much as he does with Dupin's mysteries. Poe may have inadvertently created a hoax through his publication of the story in a magazine known for news reporting as well as fiction, through his adherence to a style of writing consistent with science writing in general and medical case studies in particular, through his pretense of name-dropping, and through his fortunate exploitation of a hot topic. This second explanation accounts equally well for Poe's ambivalent reactions to readers' requests for verification.

The telling clue in this dilemma of authorial intent is Poe's reaction to readers' responses. Once Poe realized he had fooled some of the readers of M. Valdemar, he decided to own the hoax and capitalize on its publicity. This reaction is a piece to our puzzle of author intentionality in hoaxing, a piece that would not be available to us in a strictly text-based analysis of the genre. M. Valdemar counts as a hoax because *readers constructed it as such and Poe owned that construction*. Whether he intended to make a hoax when he wrote the story is irrelevant. Since the hoax is not a text, but time- and space-bound exchange with readers, Poe had many opportunities over the months of debate surrounding the story to construct himself as a hoaxer, and he indeed did. Therefore, M. Valdemar counts as a hoax. Even putting the vexed question of author intentionality aside, M. Valdemar"still adhered to popsci. expectations and appeared in a journal carrying news, so readers still had to decide on its truth-value. For the purposes of this book, it is that interpretive tension that makes M. Valdemar worthwhile as a site to examine the interaction of reading expectations about science news.

There is a fascinating dynamic operating in M. Valdemar that may help explain its complex effect on readers and that can only be explained in terms of readerly expectations. Barthes attributes the appeal of M. Valdemar to its "undecidability," to the fact that Poe invokes several competing codes simultaneously with his language in the tale—as when he invokes the enigma code and the scientific code, for example, with the phrase "the facts" but does not give the reader enough information to "choose which is 'true,'" the mystery story or the scientific report on mesmerism.[96] The story's adherence to popsci. expectations invokes the code of science. But the enigma code, with its presupposition of mystery, with its elided names and ejaculations of disbelief, constructs a puzzle instead of truth.

This confusion of two "filters" of reading expectations may explain why Poe reported receiving so many questions about the truth of the tale. His story, strung stylistically between two genres, disabled many readers from making even a provisional decision for themselves about its truth-value. Through the invocation of two conflicting codes or "filters" of reader expectations—one applicable to reports of scientific fact, and the other employed in the reading/hearing of mysteries and "ghost stories" in Elizabeth Barrett's words—Poe gave his readers a truly troubling reading experience. Poe was known for generic innovation. The craze over mesmerism in 1845 afforded him a chance to play with the curious blend of epistemologies it represented—scientific and mystical—in order to destabilize both readers' perceptions of reality and of generic conventions.

"Von Kempelen and His Discovery"

Poe's last hoax, published just months before he died, was the most calculated and highly engineered of his hoaxing attempts. The story purported to be a more personal follow-up to a series of other "scientific" reports already published in the American media about an incredible discovery by German alchemist, Von Kempelen. A lengthy preamble discussed and disparaged many of these other accounts, saying of one that it had a very "moon-hoax-y air."[97] Then, still without announcing what exactly the amazing discovery was, Poe claimed that famous chemist Sir Humphrey Davy had reported coming very close to making the same discovery in his "Diary"; an editorial comment immediately interjected that, lacking "the algebraic signs necessary, and as the 'Diary' is to be found at the Athenaeum Library, we omit here a small portion of Mr. Poe's manuscript.—ED."[98]

Poe went on to link Von Kempelen with the Maelzel family, the creators of the famous chess-playing automaton that magician David Brewster and then Poe himself debunked. Von Kempelen's supposed reputation in the American media as a misanthrope was next raised and dismissed, and then, finally, nine long paragraphs into the account, Poe reported Von Kempelen's attempts to keep his discovery secret and the subsequent raiding of his Bremen flat by police to reveal that he had changed a trunk full of lead into "gold, in fact, absolutely pure, virgin, without the slightest appreciable alloy!"[99] The account finished with the news that the price of gold was plummeting in Europe and would soon do the same in America, as a result of Von Kempelen's discovery.

Poe ostensibly wrote the hoax to take the shine off the gold rush of 1849, or so he claimed in his letter to Evert A. Duykinck of 8 March 1849, where he attempted unsuccessfully to convince Duykinck to publish "Von Kempelen" in his journal the *Literary World*:

> Dear Sir,
>
> If you have looked over the Von Kempelen article which I left with your brother, you will have fully perceived its drift. I mean it as a kind of "exercise," or experiment, in the plausible or verisimilar style. Of course, there is *not one* word of truth in it from beginning to end. I thought that such a style, applied to the gold-excitement, could not fail of effect. My sincere opinion is that nine persons out of ten (even among the best-informed) will *believe* the quiz (provided the design does not leak out before publication) and that thus, acting as a sudden, although of course a very temporary, *check* to the gold-fever, it will create a *stir* to some purpose.
>
> I had prepared the hoax for a Boston weekly called "The Flag"—where it will be quite thrown away. The proprietor will give me $15 for it on presentation to his agent here; and [page 2:] my object in referring the article to you is simply to see if you could not venture to take it for the "World." If so, I am willing to take for it $10—or, in fact, whatever you think you can afford.
>
> I believe the quiz is the first deliberate literary attempt of the kind on record. In the story of Mrs Veal, we are permitted, now & then, to perceive a tone of *banter*. In "Robinson Crusoe" the design was far more to please, or excite, than to deceive by verisimilitude, in which particular merely, Sir Ed. Seaward's narrative is the more skilful book. In my "Valdemar Case" (which *was* credited by many) I had not the slightest idea that any person should credit it as any thing more than a

> "Magazine-paper"—but here the whole strength is laid out in verisimilitude.
>
> I am very much obliged to you for your reprint of "Ulalume."
> Truly Yours,
> Edgar A Poe.
> [page 3:] P.S. If you feel the least *shy* about the article, make no hesitation in returning it, of course:—for I willingly admit that it is not a paper which every editor would like to "take the responsibility,' of printing—although merely as a contribution with a known name:—but if you decline the quiz, please *do not let out the secret.*

Poe's references to Mrs. Veal and Crusoe were to Daniel Defoe's literary adventure hoaxes, and "Sir Ed. Seaward's narrative" is a reference to a diary forged by Jane Porter published in London in 1831.[100]

Bernard Pollin takes Poe's high expectations for his last hoax seriously based on the success of M. Valdemar;[101] however, the claim that he could cure Americans of their gold fever might rather have been a ruse to entice Duykinck to publish the story. Daniel Hoffman argues that Poe's hoaxing was never aimed at "show[ing] men how to amend their ways," but rather in "display[ing] the follies of mankind—and the personal superiority of the Artist-Genius to the generality of fools."[102]

Von Kempelen was sold to *The Flag of Our Union* in Boston after Duykinck turned it down, where it appeared on 14 April 1849. From the beginning, the story stumbles over dropped names and borrowed authority: "After the very minute and elaborate paper by Arago, to say nothing of the summary in Silliman's Journal, with the detailed statement just published by Lieutenant Maury, it will not be supposed, of course, that in offering a few hurried remarks in reference to Von Kempelen's discovery, I have any design to look at the subject from a scientific point of view."[103] Dominique Arago was a renowned French physicist, head of the Observatoire de Paris at the time of Poe's hoax; interestingly, Arago had been one of the first European scientists to publicly denounce Locke's Moon Hoax in 1835 on behalf of his friend Sir John Herschel. "Silliman's Journal," the *American Journal of Science*, has already been discussed in some detail, and Lieutenant Maury was a navy engineer responsible for great improvements in American navigation before the Civil War. Curiously, he was also at the helm of the *Southern Literary Messenger* for a few years in the 1840s after Poe had left the paper.

Poe seems bent on fulfilling his readers' foreign and wisdom expectations in this hoax, along with his usual obsession with detail. He lifts whole sections of chemist Sir Humphrey Davy's memoirs and claims reports of Von Kempelen's discovery and personality have already appeared in no fewer than three major papers of the time, the *Courier and Enquirer*, the *Home Journal*, and Duykinck's *Literary World*. The details from Davy's diary, which would ostensibly satisfy readers' detail expectation, are in fact abstracted, fragmentary, and would likely have seemed nearly impenetrable and distracting to Poe's readers: "But to return to the 'Diary' of Sir Humphrey Davy. This pamphlet was not designed for the public eye. . . . At page 13, for example, near the middle, we read, in reference to his researches about the protoxide of azote; 'In less than half a minute the respiration being continued, diminished gradually and *were* succeeded by analogous to gentle pressure on all the muscles.'"[104] The story is chock full of these sorts of incoherent "samplings" of institutions, journals, or individuals Poe must have felt his reader would recognize and place confidence in. He sticks tight to the foreign expectation by using Von Kempelen's name and nationality to tap into the superior reputation of German science.[105] To conform to his readers' sensation expectation, Poe winds his story up with the sensational police raid of Von Kempelen's flat and the report of the plummeting gold market in Europe. The sagging markets, in fact, appeal to the use expectation by showing readers the immediate payoff (although negative) of Von Kempelen's discovery.

Thus, on the surface, Von Kempelen appears to meet most of the expectations identified in this book, with the exception of popsci. subexpectations long (since no further news was promised), decoration (decorative woodcuts), and analogy (explanation of phenomenon with reference to "real-life" experiences); however, these expectations are never debated by readers. Oft-debated constraints such as sensation, novelty, and detail are (to our eyes) satisfied by Von Kempelen. *The Flag of Our Union* did report science news and political news, so Poe's hoax had a fair chance at being bolstered by its readers' medium expectations. We would therefore expect the hoax to have been successful with its readers.

In fact, however, there is no recorded reaction to Von Kempelen whatsoever, and while Poe biographer Arthur Hobson Quinn claims that the hoax was one of the most "successful attempts of Poe to imitate a science report,"[106] other scholars seem to agree that Von Kempelen did not make anything like the stir Poe had intended. There could be many reasons for this, some of which may simply be due to poor recording: issues of *The Flag of Our Union* for several years surrounding 1849 are missing, and Poe does not discuss the story in his letters. He wrote few letters, anyway, between April and his death six months later in October, and they were mostly frantic pleas for money and comfort. The success of "Von Kempelen" as a hoax was simply not a priority with him at that time.

However, assuming that we have all the reception data we need, and therefore that the public simply failed to fall for Von Kempelen, can the reader expectations we have collected help explain the failure of Poe's final, and most deliberately crafted, hoax? In fact, given the lack of reception information, the common reader expectations we have collected throughout an examination of Poe's and Locke's hoaxes so far are the only chance we have to explain the failure of Von Kempelen.

The authority expectation states that the previous reputation of the author holds, and Poe's readers would, by 1849, associate him with the Balloon-Hoax at least, if not also M. Valdemar, Hans Phaall, and the public debate over the Moon-Hoax. It did not help his chances for ending up on the right side of the authority expectation that he made mention in Von Kempelen not only of the Moon Hoax, but also the Maelzel exposé, which, even if readers did not know he had forged it, would still strengthen the tie between "Poe" and "hoax" in their minds. So the authority expectation would actually work against him in this hoax.

What, then, can be said for the incredible weight of foreign names and domestic sources Poe used to keep Von Kemplen ballasted in reality? Bernard Pollin is of the opinion that Poe actually shot himself in the foot with his slavish attention to the wisdom and foreign expectations: "[Von Kempelen] is, indeed, a 'tired' kind of hoax, which defeats its purpose by presenting too much of the familiar from which readers could check on its authenticity."[107] Poe's overboard name-dropping might well have made his story ring a bit off-key, as popular science reports usually sacrificed extensive citation to keep up the excited pace of discovery. It appears Poe might have employed so much detail in this hoax—with much of it fragmentary and random, violating expectations of internal coherence—that he bogged down the sensational element. Overall, Poe made a misguided guess that his readers would rank foreign, wisdom, and detail over sensation.

It is also likely that Poe's hoax violated novelty and plausibility in ways that are not apparent to us as anachronistic readers. While alchemy was certainly the fodder for popular fiction in the nineteenth century—the Rosicrucian novel *St. Leon* by William Godwin being the stand-out example of this subgenre[108]—the "pseudoscience" sustained no discussion in popular science journals and was probably counted too medieval to have a hold on public faith. Poe's seemingly unerring sense of what "wowed" the public—such as aviation, automatons, exploration, cryptography, and hypnotism—seems to have gone awry in this final hoax.

In conclusion, three important observations can be made about the interaction of reader expectations in the hoax-reading experience as a result of the failure of Von Kempelen.

- Novelty and Plausibility may be consistently ranked above popsci., contrary to Poe's ranking, which claims Popsci. or "verisimilitude" is more important than Plausibility. Von Kempelen's failure shows us that if alchemy is simply not a novel or plausible scientific topic, it will not matter to the reader how close its presentation mimics a "true" science report. This indicates a general ranking of experience with the topic over trust in rhetorical form.

- Expectations are not met or violated in isolation, but in interaction with other expectations. Poe had a great deal of detail in this story. But since detail appears to compete with sensation in popular science reading, too much detail retards sensation. Expectations can fail by being overmet just as well as by being undermet, depending upon their interactions with other expectations.
- Poe interpolated a great deal of "tangential" information between the popsci. elements. In many places, the tone of Von Kempelen is chatty and gossipy, more typical of one of Poe's Literati portraits than a news story.[109] Interpolation of material from other genres, as seen in the case of M. Valdemar, can confuse readers and make it difficult for them to play decisive interpretive games. Poe's nonlinear personal commentary in Von Kempelen may have obscured the popsci. elements for his readers.

When viewed from the perspective of common reading expectations, Poe's final hoax appears to have failed because, ironically, he tried too hard. In a concentrated attempt to mimic all the features he believed to drive a successful hoax, he overloaded the story with fragmentary details, citations, and comments that blurred its structure and confused its readers.

Solutions to Problems in Poe Scholarship

In this book I am redefining hoaxing as a metagenre, a game played between author and reader in a news medium over readers' expectations about a particular genre, science news, in this case. This approach to Poe's hoaxing revises previous work on it in two important ways: it accounts for the multiple generic classifications of Poe's hoaxes over the years, and it restrains the overapplication of the term *hoax* that has plagued Poe scholarship since the 1960s.

First, acknowledging that we must theorize reader expectations to understand hoaxing helps account for the confusion over how to classify Poe's hoaxes since their original printings. For example, we are now prepared to explain why M. Valdemar is classified both as a hoax—by Poe's contemporaries and by current science studies scholars such as Alexander Boese—and as a science fiction tale by critics such as Roland Barthes and Bruce Franklin. The generic ambiguity is not a strictly textual function but resides in shifting reader expectations about medium and context.

Changing the medium of publication changes the criteria by which the reader assesses the truth-value of the story. When a reader encounters M. Valdemar in a literary collection, it is most likely that she suspends judgments on truth-values altogether, since those are not decisions pertinent to fiction reading. However, when the original readers of M. Valdemar encountered the story in *The American Review*, which regularly carried political and science news, decisions about truth figured centrally in their interpretation of the text.

Shifts in context have occurred as the society in which Poe's readers lived has changed. George Kennedy describes this transformation through the lens of the interpretation of classical rhetoric texts throughout European history in *Classical Rhetoric and Its Christian and Secular Tradition from Ancient to Modern Times*. *Letteraturizzazione* is the process by which a rhetorical text comes to be read under literary, or "decoupled," genre expectations. Kennedy tracks the progress of Cicero's *De Oratore* and Longinus's *On the Sublime* from their Roman reception as technical manuals for the production of political discourse, to their nineteenth-century Belletristic employment as catalogs of tropes and rhetorical devices to be reproduced in literary writings divorced from action in the public sphere. Many forces can drive this shift in reading expectations, but primary among them is the outmoding of the text's original arena of application—a democratic assembly, in the case of Cicero's and Longinus's texts.[110]

In the case of M. Valdemar, to take one of Poe's hoaxes for comparison, removing it from its original news medium and from a heated kairos of debate over mesmerism's scientific potential forces generic reevaluation. Readers are incapable of taking mesmerism as seriously ten years after the hoax's original publication, much less 150 years later. M. Valdemar has undergone *letturaturizzazione*, has become science fiction by default because its topic is outmoded in the modern reading context, and it now appears in literary media rather than news media; thus, it has lost its ability to affect readers' perceptions of reality. Any analysis of Poe's hoaxes that ignores the reader's expectations about medium and context in assigning a text to a genre will miss this crucial point.

It is hard to concretely illustrate the *letturaturizzazione* of M. Valdemar because of the difficulty of producing diachronic readings of the story. However, comparison of the hoax with a near-contemporaneous piece of science fiction may be helpful in illustrating the transformations that a literary context can effect on arguments about scientific reality. Writer Fitz-James O'Brien came to New York in 1852 a few years after Poe's death and published science fiction stories in literary magazines such as *Harper's New Monthly* and *Atlantic Monthly* until he died fighting for the Union in 1862. His stories were on topics remarkably similar to Poe's hoaxes: "How I Overcame My Gravity" recounted the experience of someone who flies into the atmosphere with the aid of a gyroscope, and "The Bohemian" concerned the gold rush and mesmerism, as did M. Valdemar and Von Kempelen.[111] O'Brien's stories, however, were never called hoaxes, likely because they were never framed in a news context, which several of Poe's were. Even in the borderline case of Hans Phaall, which appeared not in a newspaper but in the *Southern Literary Messenger*, our reconstructed reader expectations are still sufficient to distinguish Poe's hoax from O'Brien's science fiction.

O'Brien's "How I Overcame My Gravity" appeared in *Harper's New Monthly*, which began publishing in 1850, the year after Poe's death. It is on a topic very close to Hans Phaall: in the story, a scientific dabbler manages to fly high into the atmosphere with the aid of a gyroscope. Both stories are contained in literary monthlies, though because of its lack of competition, the *Southern Literary Messenger* was forced to carry a great deal more political and general news than *Harper's* was, so Poe's hoax had that advantage. However, compare the openings of the two stories:

> "How I Overcame My Gravity"
>
> I have all my life been dallying with science. I have coquetted with electricity, and had a serious flirtation with pneumatics. I have never discovered any thing, nevertheless I am continually experimentalizing. My chambers are like the Hall of Physics in a University. Air-pumps, pendulums, prisms, galvanic batteries, horse-shoe magnets with big weights continually suspended to them: in short, all the paraphernalia of a modern man of science are strewn here and there, or stowed away on shelves, much to the disgust of the maid-servant, who on cleaning-day longs to enter the sanctuary, yet dare not trust her broom amidst such brittle furniture.[112]
>
> "Hans Phaall—A Tale"
>
> By late accounts from Rotterdam that city seems to be in a singularly high state of philosophical excitement. Indeed phenomena have there occurred of a nature so completely unexpected, so entirely novel, so utterly at variance with pre-conceived opinions, as to leave no doubt on my mind that long ere this all Europe is in an uproar, all Physics in a ferment, all Dynamics and Astronomy together by the ears.[113]

The difference between the rhetorics of these openings seems intuitive. Poe's seems much more serious and newsy than O'Brien's. The difference is actually attributable to the mystery expectation. Poe is introducing an "unexpected" "novel" discovery. O'Brien is rather humorously setting the stage for a personal account of misadventures in science. Readers would not be likely to confuse O'Brien's rhetoric for that of a science news report, especially as in its original print context it immediately followed an extraordinarily sappy love story entitled "Cool Captain." At least Hans Phaall had the good fortune, for its hoaxing project, to share the page with a nonfiction piece, a critical history of English poetry. This simple comparison of the rhetoric and immediate print context of O'Brien's science fiction with Poe's hoax does not settle the issue of the confused classification of Poe's hoaxes, but it offers additional evidence that reader expectations must be consulted when assigning texts to genres.

The other major contribution a reader-expectation-based methodology makes to the conversation about Poe's hoaxing is to help salvage that activity as something special and significant as compared to Poe's other fiction practices. Beginning in the 1960s, there was a strong tendency in Poe scholarship to recategorize most, if not all, of his tales as "hoaxes." In the vanguard of this trend was Richard Benton's "Is Poe's 'The Assignation' a Hoax?" in 1963 followed by G. R. Thompson's "Is Poe's 'A Tale of the Ragged Mountains' a Hoax?" and a slew of other analyses claiming hoax status for "The Mystery of Marie Rogêt," "The System of Dr. Tarr and Prof. Fether," "The Murders in the Rue Morgue," "The Premature Burial," and *Eureka*, among other texts.[114] Marie-Louise Nickerson Matthew, in her 1975 dissertation "Forms of the Hoax in the Tales of Edgar Allan Poe," finds that all of his tales are hoaxes—either external hoaxes fooling readers or internal hoaxes giving Poe himself provisional illusions of epistemological stability. Published a few years after that analysis, the essays in Dennis Eddings's 1983 collection *The Naiad Voice: Essays on Poe's Satiric Hoaxing* mark the height of this fashion of hoax hunting.

This trend was ostensibly well motivated, as Poe actually used the word *hoax* in reference to his *jeux d'esprit* in the news media; in addition, his predilection for codes, cryptograms, and other forms of "mystification" was legendary. Problematic for these reanalyses of Poe's tales, however, is their tendency to assign the same story indiscriminately to several related genres. For example, G. R. Thompson refers to tales such as "The Assignation" as "hoaxlike parodies,"[115] and Benjamin Franklin Fisher variously categorizes the rhetoric of "Tarr and Fether" as "hoaxing," "self-parody," "satiric," and "burlesque" all in a single page.[116] Some of this generic confusion can be easily clarified through more rigorous attention to the special effects of parody, satire, and burlesque on readers, as demonstrated in chapter 1. However, in addition, viewing hoaxing as a special game in guessing and satisfying reader expectations provides a powerful tool for focusing the application of 'hoax' correctly and precisely in Poe criticism. To illustrate this point we can revise from a rhetorical perspective two of the more recent reclassifications of Poe's fiction as hoaxes.

John Bryant, in his study of "Murders in the Rue Morgue," determines that the story is in reality a hoax. Bryant cites the sociology of Johan Huizinga and Clifford Geertz in a definition of Poe's hoaxing practice as a "satiric antiritual" that in its mean-spiritedness denies its readers the comic closure of being able to laugh at themselves and thus to release the tension of the author's attack on them.[117] This analysis jibes with the picture of hoaxing we have been developing in many ways. However, in illustrating the "satiric antiritual" of Poe's hoaxing, Bryant does not choose one of the media hoaxes. Instead, he focuses on what is widely considered to be the first detective story. Bryant claims that "Rue Morgue" is a hoax because Poe hides the clues to the L'Esplanayes' deaths so well that the reader cannot figure them out and must defer to Dupin's genius and his eleventh-hour revelation of the clump of orangutan hair that clinches everything. Thus, hoaxing, in Bryant's analysis of "Rue Morgue," is simply not playing straight with the reader.

While "Rue Morgue" and the media hoaxes indeed share a theme of science—the newly developing field of forensic science, in the case of "Rue Morgue"—several dissimilarities between the rhetoric of the two practices, and the conditions of their publication, suggest that Bryant's crying hoax over "Rue Morgue" is premature and threatens to efface a rich and important distinction in Poe's rhetorical practices. Poe was very aware of the rhetorical game he was playing with readers in his detective tales such as "Rue Morgue," and it was a game quite different from the one he played in his media hoaxes. He intended the secret machinery of his hoaxes to remain concealed for the duration of the reading, for at least some of his readers. By contrast, in his detective fiction, Poe admitted to having "woven" highly artificial mysteries that he would then set about "unraveling" before the reader's eye; this became such a standard rhetorical procedure for him, in fact, that he grew weary of it and openly burlesqued himself doing it in "Thou Art the Man!"[118] So, the first and most obvious response to Bryant's hypothesis about the hoax status of "Rue Morgue" is Occam's Razor: what do we have to gain by reclassifying "Rue Morgue" as a hoax when Poe has already identified it as a genre of mystery making distinct from the parasitic metarhetoric of the science hoax? Bryant would perhaps argue that acknowledging "Rue Morgue" as a hoax uniquely reveals the "satiric antiritual" Poe puts his readership through to his benefit and their shame. But it is hard to see how a reader encountering "Rue Morgue" in the very literary *Graham's Magazine* in 1841 would experience that tale in the same way as Locke's Moon Hoax or Poe's Balloon-Hoax in the New York *Sun*; the medium expectation would serve to classify "Rue Morgue" as literary fiction rather than news. Admittedly, the opening of "Rue Morgue" does use a few words that would fit the mystery expectations of science newsreaders, such as *glories* and *enigmas*. However, they do not describe a new discovery, but rather abstract psychological concepts:

> The mental features discoursed of as the analytical, are, in themselves, but little susceptible of analysis. We appreciate them only in their effects. We know of them, among other things, that they are always to their possessor, when inordinately possessed, a source of the liveliest enjoyment. As the strong man exults in his physical ability, delighting in such exercises as call his muscles into action, so glories the analyst in that moral activity which disentangles. He derives pleasure from even the most trivial occupations bringing his talents into play. He is fond of enigmas, of conundrums, of hieroglyphics; exhibiting in his solutions of each a degree of acumen which appears to the ordinary apprehension preternatural.[119]

This opening is very nonnewslike in its musing abstraction and its failure to lay claim to the witness of a spectacular new scientific or technological phenomenon. Already the reader is alerted that the interpretive decisions they must make here will have little to do with establishing the truth-value of the story to follow. The immediate failure of "Rue Morgue" to satisfy the crucial reader expectations of popsci. and medium about science news ensures that the story will not be read as fact, but as fiction. How, then, can Poe target the readers of "Rue Morgue" to embarrass them for their dimwittedness, if they are not prepared to make interpretive decisions that engage their assumptions about reality? The readers of "Rue Morgue" are likely practicing "willing suspension of disbelief," to borrow Coleridge's original description in the *Biographia Literaria* of the "decoupling" of meaning from world view that is the hallmark of fiction.[120]

Probably the most broad-based attempt to reclassify Poe's fiction as hoaxing was Marie-Louise Nickerson Matthew's 1975 dissertation "Forms of Hoax in the Tales of Edgar Allan Poe." As mentioned briefly above, Matthew claimed that all of Poe's fiction was hoaxing, either external hoaxes to dupe readers or internal hoaxes to provide Poe himself with fantasies that provisionally stabilized his erratic mind.[121] Matthew's definition of hoax fails at a high level, not at a more local level like Benton's and Bryant's. Her analysis elides the hoax's primary function of identifying and transforming reader expectations; this crucial mistake is what allows her to overapply the term. Matthew ignores the fact that Poe's hoaxes transform reader expectations about genre and about the world. This transformation is accomplished through revelation of "the truth," which runs counter to the argument of the hoax and thus forces readers to reexamine their assumptions. Either Poe's hoaxes reveal themselves during the reading process, for cannier readers, or Poe himself reveals them, for the less canny. No one stays in the dark, or it is not a hoax (*hoax* again denotes not strictly the text but the whole game over truth between an author and a reader through a news medium). A hoax's very raison d'etre is to undermine expectations. Where, then, is even the "provisional" stability that Matthew claims that the "internal" hoaxes like "Murders in the Rue Morgue" provided for Poe? Hoaxes are in the business of destabilizing reality, not stabilizing it.

The preceding are three examples of how attention to reader expectations about genre can help prevent overapplication of 'hoax' to all of Poe's work. However, consideration of these problems has in turn provided a crucial contribution to this project: a provisional overall ranking of the reader expectations collected so far. The fact that readers read Poe's hoaxes as hoaxes when they are in newspapers but as science fiction or detective fiction when they are in literary collections indicates that the expectations of medium and authority dominate other reader expectations. These high-level expectations determine the kind of interpretive game that will be played from there on in. If medium and authority support factual news-reading, then games over truth will be played, utilizing the expectations about the "real world" and the rhetoric of science news collected in this project. If medium and authority support a fiction reading game, then the entertainment expectation kicks in, the game is "decoupled" from decisions about truth, and other important literary subexpectations under Entertainment apply, which are beyond the scope of this book.[122] Checking this hypothesis against the reader responses we have tabulated so far results in confirmation. Thus, I propose the following common reading filter for antebellum science news:

{Corroboration, Medium, Authority}>>
{Novelty, Sensation, Plausibility}>>
{Popsci., Foreign, Internal Coh.}

Again, ranking is from right to left, lowest ranked to highest ranked. Double angle brackets ">>" denote levels of rank, as would bold vertical lines in the tabular notation. Commas, like the dotted lines in the tabular notation, indicate lack of evidence for competition, and therefore equality of strength, among expectations in a given level of rank.

Starting at the lowest level of rank, foreign and internal coherence were relatively local expectations that were both related to and therefore seemed to have the same strength as popsci. (verisimilitude) expectations. More important to most readers were novelty, sensation, and plausibility. Finally, expectations about medium and authority are highest ranked, "trumping" any other expectations in case of conflict with them. Although not as frequently mentioned by readers, corroboration, intuitively linked with the activity of checking the reputation of sources, never lost an interpretive game it was involved in, so it resides on the highest rank. Only when these highest-level expectations have been satisfied or obviated due to absence of information about author and medium (as in a newspaper clipping) are the lower-level ones allowed to play decisive roles in decisions about truth.

This provisional ranking represents the filter the majority of Poe's readers may have been (unconsciously) using when approaching his hoaxes. It is unlikely that all of the expectations would have figured in any single reader's interpretation of one of Poe's hoaxes; reader decisions, as we have seen, tend to focus on just a few competing expectations, even if others are in play unconsciously. The filter, then, synthesizes the expectations and ranking information gleaned from twenty-two extant individual reading decisions about Poe's and Locke's hoaxes. It therefore represents in skeleton form part of Poe's readers' world view, their beliefs about science and science news—the ideational place where Poe engaged them and built a public relationship with them through his hoaxing.

Poe's Relationship to Science and to His Readership: How the Hoaxes Interact with *Eureka*

Now that we have examined how Poe fooled his readers with his hoaxes, the more interesting questions remain: What was he after, and why? These questions both come into focus when we reflect his hoaxing practices off *Eureka*, Poe's fullest statement of the relationship of the writer to reality and to his or her readership.

Eureka was published in 1848, between the hoaxes M. Valdemar and "Von Kempelen." There is ample evidence that Poe felt *Eureka* would be the crowning achievement of his literary career, from the excited letters he wrote friends and publishers about the book. It represented an astounding range of scientific and philosophical reading. John Limon finds that the book's clearest debt is to the German school of *Naturphilosophie*, specifically to Schelling and Hegel, whose works had been published in English in America just a few years before the publication of *Eureka*.[123]

It is actually easy to put the point of *Eureka* briefly because it was an argument that the entire universe was constructed and sustained by two and only two opposing forces, gravitation and electricity, called elsewhere "attraction" and "repulsion," and "Unity" and "difference."[124] Poe anticipates deconstruction by claiming the universe is always in the process of its own undoing: "My general proposition, then, is this:—In the Original Unity of the First Thing lies the Secondary Cause of All Things, with the Germ of their Inevitable Annihiliation."[125] In a slightly ironic twist of fate, Poe was indebted to Richard Adams Locke for this idea. At the age of eighteen, Locke had written a poem in six cantos about the cyclic destruction and rebirth of the universe, and at a Lyceum lecture he gave in Boston on magnetism, he expounded a refined version of this theory. Poe availed himself of the pamphlet and the concept for *Eureka*.[126]

Poe begins the body of argument in *Eureka* with what has been called a "hoax," but what is really a bizarre recap of "Mellonta Tauta," where a letter fallen from a balloon time traveling from the year 2848 criticizes the nineteenth century for its scientific backwardness. The old syllogistic system of "a Turkish philosopher called Aries and surnamed Tottle" is ridiculed for its rigidity, as is the "crawling" inductive system of "one Hog surnamed, 'the Ettrick shepherd'" (Francis Bacon).[127] The only way to truth, implies Poe through the tinny voice of the letter writer, is through the imagination:

> [Y]ou can understand how restrictions so absurd on their very face must have operated, in those days, to retard the progress of true Science, which makes its most important advances, as all History will show, by seemingly intuitive *leaps*. These ancient ideas confined investigation to crawling; and I need not suggest to you that crawling, among varieties of locomotion, is a very capital thing of its kind; but because the snail is sure of foot, for this reason must we clip the wings of the eagles?[128]

The letter writer goes on to complain that the Baconian school of philosophy demanded that every truth be demonstrated empirically. Because true demonstration of anything is impossible due to the essentially individual and contingent nature of observation and belief, this unjust requirement stultified the growth of science, and "No man dared utter a truth for which he felt himself indebted to his soul alone."[129]

On the contrary, Poe says when he is done ventriloquizing, to understand a concept as awesome and indemonstrable as the nature of the very universe, a different methodology must be developed. His new science sounds in places a great deal like Dupin's intuitive ratiocination. Here, Poe describes it as a "whirling on the heel" on the top of a mountain in order to take in the whole panorama in one sublime blur.[130]

Of course, the rest of his argument about how gravitation and electricity work in the universe smacks a great deal of both Aristotelian syllogism in some places and Baconian induction in others. Although he claims that it is impossible to prove anything "axiomatically," he often works through syllogisms to build one part of his argument on another. Moreover, he seems to have a compulsive need to measure his ideas against the "real world." But Poe is actually attempting a great experiment in a slightly different system of reasoning here—analogy, through which truth is determined on the basis of correspondence of unknown phenomena to known phenomena, like the radiation of light and heat.[131] Poe writes that "a perfect consistency can be nothing but an absolute truth,"[132] and as he does, he indebts himself to Francis Wayland's conception of analogical reasoning, as well as to Humboldt's *Cosmos* and to Laplace's nebular theory.[133]

It has often been ventured, particularly by critics who have chosen to view him as a romantic, that Poe hated science. Certainly, a cursory reading of "Sonnet—To Science" and of his catty comments about Bacon in *Eureka* and elsewhere lends credence to this notion. But a careful examination of Poe's hoaxes and *Eureka* show that Poe was fascinated with science. He just favored his own science of imaginative leaps over what he viewed as the baby stepping of Baconian induction. Further, he was not against all modern induction, just the version he believed to be common currency in America. His disdain for this plodding sample collection with no inductive speculation is apparent in a letter for the *Columbia Spy* where he reviews the Wilkes expedition.[134] He calls it "encumbered with 'men of science.' Let some Yankee open the way (as, assuredly, some Yankee yet will), and let men of science follow his footsteps, and geologize at their leisure."[135] In *Eureka* Poe again champions the adventurer with imagination over professional Baconian scientists.

> [M]erely perceptive men . . . those inter-Tritonic minnows, the microscopical savants, the diggers and pedlers of minute *facts*, for the most part in physical science; *facts*, all of which they retailed at the same price upon the highway; their value depending, it was supposed, simply upon the *fact of their fact*, without reference to their applicability or inapplicability in the development of those ultimate and only legitimate facts, called Law.[136]

Poe's hoaxing dovetails neatly with this credo about the imagination. His hoaxes, as already pointed out, were designed to embarrass readers, but for what, and to what end? Daniel Hoffman has already argued convincingly that Poe was not interested in helping any of his readers learn more about science. With his hoaxes, Poe was engaged in a campaign to embarrass people for letting the rigid limits of modern science blind them to the truth they could have apprehended if they had used their intuition. Thus, Poe points out his superior scientific-imaginative genius in being able to connive rhetorical contraptions that will dupe the reading public. The hoaxes, seen in this light, become an indirect argument, an advertisement, almost, for the transcendence of Poe's intuitive ratiocination over Baconian induction or Aristotelian deduction.

In the discussion of Poe's Balloon-Hoax, the mechanistic nature of both the hoax and Poe's rhetoric was highlighted. I argued there that the hoax was actually a sort of machine or automaton that took in reader expectations and transformed them into an experience of shame or embarrassment. Daniel Hoffman has found the same mechanical aesthetic at work in *Eureka*. By boiling down the entire universe into two inexorable and completely balanced forces, Hoffman argues that Poe shouts to his readers, "The entire universe is a huge coherent contraption!"[137] Further, Hoffman contends, Poe proves through *Eureka*, through the *philosophy of composition*, the *poetic principle*, and other writings that he is essentially a "mechanician of literature and his theories a program for the production of verbal contraptions."[138] The hoaxes are some of these "verbal contraptions": Poe carefully built them, tinkered with them, watched their progress in the world with the worried and elated anxiety of an inventor. They are designed to transform readers. When they work, they suck readers in and then, after spitting them out, leave them to look over the gears and pistons and marvel at the genius of the man who was able to do this to them.

This leads us to the question of what kind of relationship Poe wanted with his readership, because humiliation does not endear. Most analyses leave the relationship where we just did: traditionally, Poe feels "contempt" for his reading audience.[139] He is the insecure genius who uses his hoaxes to humiliate his readership so he can feel superior. Poe's own behavior and writings admittedly support this easy dichotomy of "hoaxers" and "hoaxees." In addition to his low opinion of the "readily gullible" public expressed in his Literati note on Richard Adams Locke, Poe elsewhere mocks the public as "believers in every thing Odd," whose "Credulity:—let us call it Insanity at once," marks them as "ignorant people."[140] And why else would he throw a monkey wrench in the potentially lucrative run on copies of his Balloon-Hoax by announcing it was a hoax, unless he wanted more than money (and he wanted money!), unless he wanted the face-to-face experience of forcing a crowd of his readers to admit their gullibility and his creative authority.

But this one-dimensional view of Poe's complex literary behavior in his hoaxes is dangerously reductionist. In an alternative view outlined in his *Reading at the Social Limit*, Jonathan Elmer attributes Poe's difficult relationship with the public to an essential incompatibility in American democracy between individuality and social assimilation. Elmer argues that, for a society that champions the individual, it is ironic that both the "we" of American society and the "I" of the individual cannot be held in view at the same time because each implies the absence of the other. "'I' become riven by my participation in the social whole, and 'it' becomes riven by my exemption from it."[141] This dynamic is exactly reflected in the tension between attraction and repulsion in *Eureka*. As Poe points out in the preface, this tension in the universe is not primarily a physical, but a spiritual, social, and poetic principle as well. In spiritual terms the philosopher's independent thought creates a difference that always pulls him/her against the gravitational force trying to coalesce everything in a primordial state of Unity. And even though Poe tries to reassure the ardent individualist that it will not really sting to lose one's identity and get sucked into the One, he does not seem to buy his own rhetoric: "The utter impossibility of any one's soul feeling itself inferior to another; the intense, overwhelming dissatisfaction and rebellion at the thought; these . . . are, to my mind at least, a species of proof far surpassing what Man terms demonstration, that no one soul is inferior to another; that nothing is, or can be, superior to one soul; that each soul is, in part, its own God—its own creator."[142]

The spiritual principle is fast becoming a social principle here. For in true Heisenbergian form, the closer Poe gets to determining himself as an individual, the more isolated he feels from his community; conversely, the more he allows himself to be absorbed into the mass of American culture, the more anxiety and uncertainty he experiences about his personal identity. Karen Roggenkamp claims this anxiety explains Poe's courting of the penny press, a journalistic forum he considered beneath him. The fame and notoriety Locke attracted with his hoax made Poe want some of it for himself. Yet the closer he got to identifying with his readers, while trying to guess their expectations, the more anxious he became that he was one of them, that he, too, was a member of the mediocre "masses" to whom he so longed to prove his superiority. He commented somewhat ironically on this dilemma by way of discussing the social effect of satire in an unpublished draft of his Literati portrait of Laughton Osborn:

> [T]hus in satirizing the people we satirize only ourselves, and can never be in a condition to sympathize with the satire.
>
> It is forgotten that no individual considers himself as one of the mass. Each person, in his own estimate, is the pivot on which all the rest of the world spins round. He may abuse the people by wholesale, and with a clear conscience so far as regards any compunction for offending any one from among the multitude of which that people is composed. Every one of the crowd will cry "encore!—give it to them, the vagabonds!—it serves them right."[143]

The hoaxes are in many ways the instantiation of Poe's insecurity about his social identity; they show him both longing to be absorbed in a community of fellow thinkers and attempting to distance and dichotomize the "multitude" to whom he feared he actually belonged. A hoax, the process of reading a hoax, materializes a double readership—the readers who fall for it, and those who catch on and read it as a coconspirator of the hoaxer rather than as his/her victim. As mentioned before, those two sets of readers actually live in different worlds formed by their beliefs. It is this second world of readers that is often overlooked in assessments of Poe's hoaxing, an audience Poe invoked, created, in fact, with the clues he left in his hoaxes for the acute observer: the goofy underlying meanings of his foreign names such as Schnellpost, the fact that Hans Phaall sets out in his balloon on April 1, and the use of the name "Kissam" as a reference to a sycophantic correspondent whom Poe's friends would have recognized immediately. Kent Ljungquist argues that this verbal play is deliberate and "central to [the] hoax, a form that establishes two audiences: those deceived by the author's ironic dissembling and those cognizant of his satiric purpose."[144]

Poe would have immediately recognized this potential for duality in the mechanics of the hoax because he was already at home with doublings and double motions in his writing. His tales contain many pairs of twin characters, reflecting back to each other complementary (and often annihilatory) characteristics: Madeline and Roderick Usher, Dupin and the Minister D, Dupin and the narrator in "Murders in the Rue Morgue."[145] *Eureka*, in fact, is a double motion, both poem and scientific treatise, and the universe in it is a constant double motion of attraction and repulsion, Unity and difference.

It is exactly this double motion of attraction and repulsion that Poe was engaging in with his readership: distancing himself from a readership "too exclusively intent on the making of money" to use their imaginations or to support artists,[146] while drawing to him those few to whom he dedicated *Eureka*, those "who love me and whom I love, to those who feel rather than to those who think, to the dreamers and those who put faith in dreams as in the only realities."[147] These people Poe deemed worthy of communion; they could join him "spinning on his heel" on the summit of Aetna and agree with him that reality was not what actually happened but what could happen. These "few gifted individuals, who kneel around the summit, beholding, face to face, the master spirit who stands upon the pinnacle,"[148] could become godlike writers of reality for American readers. Like the suffocating but strangely compelling vision of communion between the narrator and Dupin in "Murders in the Rue Morgue," Poe's hoaxes, when defined as carefully engineered rhetorical transactions with a double audience, reveal him not just yearning for community but actually designing and building it. Accordingly, even his most traditionally esoteric texts, like *Eureka* and *The Poetic Principle*, deserve reexamination not as uneven attempts at creating theory, but as complex exercises in creating publics, seeking communion. Such rereadings, although beyond the scope of the present project, promise to provide a richer appreciation of Poe's rhetorical and social behavior as a pioneer of American genres.

Chapter Four

Mark Twain and the Social Mechanics of Laughter

It seems inevitable that Twain would turn to hoaxing given his penchant for satire—expressed as early as the age of seventeen with his first published story, "The Dandy Frightening the Squatter," for the humor magazine *The Carpet Bag*. That he would choose scientific media hoaxing for the mode of his first published hoax, The Petrified Man, seems an equally natural turn for Twain. Science and technology were preoccupations of his writings and business dealings from his admiration of the World's Fair in New York in 1853 to his disastrous investment beginning in 1880 in the Paige automatic typesetting machine, which Twain was certain would revolutionize the print business.

However, until now, Twain's fascination with these topics, and with the philosophy of mechanism, has mostly been treated as a biographical vehicle through which to psychoanalyze Twain in his final depressed years as a writer and bereaved husband and father. This has been the approach of scholars such as Lawrence Berkove, Tom Burnam, Pascal Covici, Sherwood Cummings, and Hyatt Waggoner. While Covici attempts to use the hoax as a figure for Twain's late-life determinism, none of these authors look to Twain's hoaxing as a necessary component of his philosophy of science. Connecting Twain's hoaxing in The Petrified Man and the Empire City Massacre with his philosophy of science and culture in later works such as *A Connecticut Yankee in King Arthur's Court* suggests a different conclusion. Instead of ambivalence about the value of science in society, or fatalism about the increasing mechanization of human culture, Twain's hoaxing practices point to a more complicated response to science and technology in American culture. They reveal the double-edged sword of rhetoric as an instrument of social control and laughter as a complex and constructive response to that rhetoric, promoting self-determination and independence.

To build an argument about Twain's hoaxing toward that conclusion, I will first examine Twain's rhetorical and scientific acculturation, focusing especially on his experience with the tall tale and how that rhetorical activity helped lay the groundwork for the development of the media hoax. Next, I will consider in detail Twain's first and major scientific hoax, The Petrified Man, both its motivations, as stated by Twain, and the reaction to it. I will use reader reactions and Twain's characterization of them to further modify the filter of science-newsreading expectations developed in the last chapter to accommodate changes in kairos since Poe's time. Finally, I will compare the results from the study of Twain's hoaxing against his scientific thinking expressed in three of his later major fiction works dealing with science and technology: *3,000 Years among the Microbes*, *The American Claimant*, and *A Connecticut Yankee in King Arthur's Court*. Claims that Twain had no social program with his hoaxing will be reevaluated in light of these works, his early hoaxes, and evidence of his deploying laughter as an attack against the power of science in American politics.

Rhetorical Acculturation

Formal Education

If Poe received from his schooling everything he would need in terms of rhetorical and scientific training to prepare him to write science hoaxes, Twain's education presents a different picture. In a letter to his brother Orion in 1865, Twain lamented having little formal rhetorical cultivation of his native talents, "[T]hough the Almighty did His part by me—for the talent is a mighty engine when supplied with the steam of education—which I have not got, & so its pistons & cylinders & shafts move feebly & for a holiday show & are useless for any good purpose."[1] Needless to say, in light of our investigation of the connection between Twain's rhetoric and his scientific and technological thinking, it is significant to find him thinking of rhetoric in mechanical terms at about the same time he is composing his first hoaxes. However, the dramatic self-regret performed in this letter is misleading. Twain prided himself on his work-a-day background and practical self-education as a riverboat pilot, a miner, and a writer.

As a young boy, Twain attended a private school in Hannibal costing a quarter a week, and that only infrequently, whenever the fishing was no good, according to his *Autobiography*.[2] There, he learned spelling and math, recited poetry and prose selections, and was taught to explicate Bible stories.[3] He may even have been exposed to some Latin and French on the days he deigned to come to class.[4] The most vivid pictures of what his classroom life might have been like do not come from his autobiography but from his fiction—for example, the school scenes in Tom Sawyer—and from his reports on schools for the *Territorial Enterprise* in Nevada. On a visit to Miss Clapp's private school in Carson City in 1864, Twain found that the schoolroom had not changed much since the days of his formal tutelage. Neither had the form of student compositions, in which Twain claimed to recognize these features of his own schoolboy writing:

> The cutting to the bone of the subject with the very first gash, without any preliminary foolishness in the way of a gorgeous introductory; the inevitable and persevering tautology; the brief, monosyllabic sentences (beginning, as a very general thing, with the pronoun "I"); the penchant for presenting rigid, uncompromising facts for the consideration of the hearer, rather than ornamental fancies; the depending for the success of the composition upon its general merits, without tacking artificial aids to the end of it, in the shape of deductions or conclusions, or clap-trap climaxes, albeit their absence sometimes imparts to these essays the semblance of having come to an end before they were finished—of arriving at full speed at a jumping-off place and going suddenly overboard, as it were, leaving a sensation such as one feels when he stumbles without previous warning upon that infernal "To Be Continued" in the midst of a thrilling magazine story.[5]

Twain's perceptive dissection of this genre of composition indicates a facility for rhetorical analysis that doubtless served him well when he wrote in parasitic genres such as satire, parody, and hoax. But, the classroom was not instrumental in developing his rhetorical skills. Rather, his newspaper apprenticeships, beginning at the age of twelve, and his affinity with oral narrative and humorous traditions account for the bulk of Twain's rhetorical preparation for hoaxing.

Newspaper Apprenticeships

After the death of his father in 1857, Twain dropped out of school for good and was apprenticed to his brother Orion at the *Hannibal Journal*, which Orion edited. Twain's duties included setting type, proofing sheets, and occasionally composing advertisements and news items. Working at the paper acquainted Twain with the actual mechanics of newspaper production as well as the role the paper's rhetoric played in its readers' lives. At a young age, he proved himself already aware of the news medium's authority to startle people and remake their world. An early hoaxlike joke was printed on the front page of the *Journal* in 1853: "Terrible Accident! 500 Men

Killed and Missing!!" read the headline. The story went on, "We had set the above headline up, expecting (of course) to use it, but as the accident hasn't happened yet, we'll say (to be continued)."[6]

It was during this time that Twain published his first satire in *The Carpet Bag* in 1852. Then, in 1853, tired of his brother never having enough money to pay him his wages, he left for St. Louis to work principally as a typesetter on the *Evening News*. He stayed only a few months before moving on to New York City to see the World's Fair and to try his hand as a compositor for a printer. Here he embarked on a program of self-education through reading at the local printer's library in the evenings, or so he assured his mother, who undoubtedly feared for his moral life in the big city. Sometime during this period, he read the satire of Laurence Sterne, Thomas Hood, and George W. Curtis. Twain also liked Oliver Goldsmith's *Citizens of the World* and Cervantes's *Don Quixote*. In a February 1861 letter to his brother Orion, he alluded to reading Dickens. Other early reading includes Thomas Paine's *Age of Reason*, Voltaire, and William Tappan Thompson's *Major Jones's Sketches of Travel.*[7] Twain enjoyed travel books in general and Herndon's *Travels in the Amazon* in particular.[8] The influences of these writers, especially the satirists, began to emerge in the pieces Twain wrote for the *Spirit of the Times*, a sporting magazine that emphasized in the style of its articles the "distinction between the false and the real and between the pretentious and the unsophisticated"—all characteristic of oral humor genres such as the tall tale and the practical joke.[9] The satire and travel narratives he read, when combined with the frontier humor tradition and his experience with the mechanics and authority of news media, made media hoaxing a natural next step for Twain.

Frontier Humor: The Practical Joke and the Tall Tale

Probably the greatest influence on Twain's hoaxing practices was his experience with frontier humor—prefigured by his early reading of *Spirit of the Times* and actualized by his stint as a riverboat pilot and his migration to the territory of Nevada during the Civil War. This exposure marks a major difference between Poe's and Twain's hoaxing but also a significant common thread. Poe went through much of the same preparation for hoaxing as Twain did—living in the North and South with different classes of Americans, reading travel narratives, writing for newspapers. But Poe did not live in the West, while Twain did for nearly half his life (Missouri counted as the West at that time). Twain situated his hoaxes on the frontier, since that was a liminal epistemological realm for his readers. Poe, living in the urbane East, had to situate his hoaxes on other borders of American experience—Europe, space, life/death, matter/energy. While frontier humor did not have the impact on Poe that it did on Twain, the idea of the frontier insinuated itself into both of their hoaxes as the line where human knowledge became insecure and thus where a hoax was most effective. It is significant that hoaxing activity died down in New York and the East Coast at the same time it was working up a good head of steam in the West—also, that western hoaxes were about the West, not Europe or Fiji or the moon.[10] Twain's hoaxing thus responded to two shifts in American humor, according to Walter Blair and Hamlin Hill. It championed the little guy, the pioneer, over the powerful and rich, and it was essentially regional, one reaction of humor to the national stresses of the Civil War.[11] The frontier was the new locus of reality making for Americans in many significant ways, and Twain and other news writers on the frontier took advantage of that fact to demonstrate authority over their readers and over the powerful new culture of science.

Frontier humor, also often called "southwestern humor," is exemplified best by the tall tale and the practical joke. Pascal Covici characterizes these forms as follows: "If there is any one pattern basic to the humor of the Southwest it is precisely this: a character is pushed by the author into a situation in which he either exposes the pretensions of others or himself emerges as ridiculous because of his pretentious behavior."[12] The satiric or critical gist of the tall tale and the practical joke initially appears incongruous with the historical notion of the frontier as a very serious zone of danger and wonder—until the social dynamics of pioneer life are more closely considered. The frontier was indeed replete with elements beyond the control of the pioneers—native peoples, lethal plants and animals, ghastly weather. In addition to contending with this environment, pioneers had to contend with each other for resources and respectability. The old hallmarks of class and caste did not apply on the frontier, "a country without a history."[13] Other methods had to be developed to secure coveted status as a savvy insider and to ostracize outsiders.

The tall tale and practical joke emerged as rhetorical mechanisms of control along the frontier. Both rhetorical modes excited laughter, which provided much-needed detachment from—and therefore a sense of objective control over—the dangers of frontier life.[14] A good example of this use of the tall tale in Twain's work is the story from *Roughing It* of Bemis and the buffalo hunt. The band of travelers that the narrator joins gets stuck after crossing the Platte River when their stagecoach breaks down. They decide to go on a buffalo hunt while they wait for rescue by the next stage, and in the process, the passenger Bemis gets chased by a buffalo and stuck up a tree until he can be retrieved by the other passengers. To regain some face, he tells an elaborate story about the ferocity of this particular buffalo and the amazing self-restraint he demonstrated in not shooting it because if he had, he claimed, his gun was so powerful it would have killed not only the buffalo but also several of the other passengers in the hunt.[15] Not only does his tale entertain the other passengers, but it is also an attempt by Bemis to regain symbolic control over both the buffalo and the ensuing humiliation.

This illusion of control served a social function, too, with respect to outsiders on the frontier—"greenhorns" or "city slickers." The telling of the tall tale or performing of the practical joke demonstrated, on the part of the insider, a superior level of comfort with and control over elements of the frontier—wild animals, storms, vicious ruffians—that terrified and bewildered the outsiders.[16] Thus, these rhetorical modes pitted knowledgeable insiders against greenhorns and used mechanisms of deception and revelation to force the greenhorns to publicly acknowledge their outsider status, much like the revelatory and humbling mechanisms of the hoax.[17] A good example of this aspect of the tall tale is the Buncombe Trial in *Roughing It*. Buncombe, a city lawyer, comes out to a frontier town to try a case in which a house has been moved on top of another house by an avalanche; the owner of the topmost house is now laying claim to the land of the bottommost house owner and wishes Buncombe to represent him. The trial gets more and more ridiculous, and more and more exasperating for Buncombe, until it is revealed that the whole thing was an elaborate charade by the townfolk to "put one over" on the city lawyer.[18]

In this way, the tall tale and practical joke both served a leveling function. Pioneers bought enthusiastically into the Jacksonian ideal of the absolute equality of common Americans and its attendant suspicion of any kind of aristocracy or undemocratic privilege. The worst charge that could be leveled against someone was that of "social impersonation," pretending to be an insider when you were not, pretending to belong to a higher class than someone else.[19] Dan De Quille, whose hoaxes will be considered in the next chapter, once gloated in a letter to his sister that easterners he met were consistently unnerved by the westerner's lack of respect for title and social standing.[20] The tall tale and practical joke in essence performed the "bootlessness" of old eastern hierarchies and mores on the frontier.

An interesting microexample of this rhetorical dynamic is a practical joke played on Mark Twain by Artemus Ward and reported by C. C. Goodwin in his memoir about the salad days of the *Territorial Enterprise*. Ward visited Virginia City for several weeks in 1863 and went out drinking with the writers and editors for the paper. With at least three humorists in the group—Twain, Ward, and Dan De Quille—there was no lack of grandstanding as each tried to better the others with quips, stories, and jokes. Ward chose an interesting tactic to "take the stuffing" out of Twain's pretensions to literary superiority when he played a practical joke on Twain in which he defined the word *sulphurets* in three increasingly incomprehensible ways while Goodwin, De Quille, and the others all nodded in mock comprehension. Twain got increasingly flustered at his inability to keep up with Ward's condescending nonsense and was livid when the joke was revealed.[21]

The practical joke and the tall tale differ from each other in a few important ways, although Twain was a regular practitioner of both, especially during his western years. The tall tale is *told* to the outsider, while the practical joke is *performed upon* the outsider. The practical joke is usually more individual and circumstantially bound than is the tall tale, which possesses a formula that may be repeated in different circumstances to a similar effect. However, the similar functions of these rhetorical forms with respect to frontier epistemology and sociology make them natural predecessors to the scientific media hoax; in fact, Constance Rourke, in her famous study of modes of American humor, marks frontier humor as the immediate progenitor of the Moon Hoax in 1835.[22] The hoax differs from frontier humor in important ways, which Twain exploited, and these will be discussed after consideration of his first hoax, The Petrified Man. Before we can account for the rhetoric of that hoax, however, we need to consider how Twain acquired not only the "insider" knowledge of science necessary to pulling off a science hoax successfully, but also the motivation for doing it.

Scientific Acculturation

The questions about the influence of science on Mark Twain's hoaxing amount to these: How did he get interested and educated in science? And, what kind of relationship with scientific culture was he trying to establish by hoaxing his readers? Twain's scientific education was quite different from Poe's. It was informal and trade-based, developed first through riverboat piloting and mining, and later through reading and investing in inventions. Twain did not have the same bone to pick with science and scientists that Poe did, since Twain did not have a competing epistemology he wished to publish. Science impacted Twain's practices at the level of authority. Scientists were competitive authors in creating the West for readers and in taking credit for that creation. During Twain's tenure in the West, the United States launched myriad scientific expeditions, like Powell's Grand Canyon expeditions and other sorties of the United States Exploring Expedition. The newspapers and publishing houses were full of their reports, which effectively created *ex nihilo* these regions for readers.[23] Often the reports were exaggerated. Twain must have sensed competition with his own exaggerated stylings of the West, because he set out to "kill" his readers' overblown opinion of the authority of paleontologists with The Petrified Man hoax. The hoax was Twain's way of cutting competing scientific authors and their "stories" down to size, while reestablishing his authority over his readership as the one writer who could deliver the authentic West.

Scholars have gone back and forth on Twain's scientific education. Certainly, science figured centrally in his reading. Albert Bigelow Paine quotes him in his *Biography* as saying, "I like history, biography, travels, curious facts and strange happenings, and science."[24] In his *Autobiography* Twain describes learning the basic tenets of evolution from one McFarlane, a well-read boarding-house laborer, several years before Darwin's *Origin of Species* was published.[25] Throughout Twain's life, he evinced interest in sciences other than just biology: geology in particular—due to his mining experience—as well as anthropology and astronomy. Post-1870 we have evidence of him reading Thomas H. Huxley's *Evolution and Ethics*, Darwin's *Descent of Man*, Bayne's *Pith of Astronomy*, and the writings of physicist and philosopher Sir Oliver Lodge.[26] Twain also showed some fascination with phrenology, palmistry, "dream science," and telepathy, although in later years he dismissed almost all of these as bunk.[27]

Twain had no formal scientific education, however, and this shortcoming has sparked the debate over his "insider" status with respect to science and technology. The prevailing opinion until recently, according to Judith Yaross Lee's reconstruction of the debate, was established in 1937 by Hyatt Waggoner. Waggoner argued that Twain's lifelong reading in the sciences and his technical experience as a printer, pilot, miner, and inventor qualified him with a level of scientific expertise in spite of his lack of formal education.[28] Lee, however, takes issue with the Waggoner school of reading Twain's scientific expertise. She points out that scientific knowledge was becoming increasingly specialized and academic in the 1860s and that Twain's casual reading in scientific books written for consumption by a popular audience did not gain him entrance into the scientific community; she figures Twain for an outsider, not an insider, and backs up her argument by comparing the humor of educated scientific experts such as West Point engineer George Derby (a.k.a. the madcap astronomer "John Phoenix") with amateurs such as Twain: "Writers with technical expertise tend to parody scientific discourse, play with scientific ideas, or experiment with science fiction. Their humor may debunk individual scientists or projects, but learning itself retains its positive value. By contrast, humorists without technical backgrounds—that is, amateurs—tend to ridicule science and the scientist as one."[29] Lee puts Twain in the latter category, citing stories of his such as "How the Animals of the Woods Sent Out a Scientific Expedition," where scientists are figured as insects who redraw the world's latitudes based on their finding a set of train tracks. Both science and scientists are made to appear woefully shortsighted and inadequate in this sketch.[30]

Part of the disagreement about Twain's scientific knowledge is due to Lee's framing of the debate. She reads Waggoner as claiming scientific expertise for Twain, while what Waggoner really wrote was this: "A study of the Notebook, the Letters, the Autobiography, the official Biography, and several unpublished sketches, discloses a knowledge of science that, while not profound or in any sense rigorously accurate, was nevertheless inspired by enthusiastic interest, and was, for the average layman of the day, comparatively comprehensive."[31] This claim is a far cry from championing Twain as a lay scientist.

In fact, Waggoner is close to expressing an observation that helps resolve the issue of Twain's scientific experience: while Twain's formal scientific training was indeed nonexistent, his technical knowledge was impressive. Even in the midnineteenth century, there was a distinction between scientific and technical practice. Scientists and engineers recognized a difference in each other's methods, which boiled down to the use of theory: scientists used it; engineers did not.[32] Engineers employed (and still do to an extent) an instinctive, hands-on, "tinkering" approach to invention that used feedback from real work environments to direct changes and improvements to technology. Engineers, mechanics, and inventors were rarely college educated, unlike most scientists of the time.

Twain's experience of science could in fact be better characterized as engineering knowledge. He gained extensive technical and mechanical know-how from his work as a printer's apprentice and as a riverboat pilot. A famous passage in *Life on the Mississippi* describes Twain's regretful transformation from a neophyte worshipper of the Mississippi to an expert pilot of the river. He claims that he came to resent the fateful sunset during which he looked at the river and could no longer enjoy its beauty as he had in his youth because it was now a complicated, technical map of shoals, submerged logs, and shallows.[33] Thus, Twain evinced an awareness of having crossed from novice to expert status as a "river technician."

When Twain headed west with his brother Orion to try his hand at mining in 1861, he continued his technical self-education with experience in mine engineering and metallurgy. In his later years, he turned to inventing, officially registering three patents (a board game, a self-pasting scrapbook, and "garment straps"). Finally, his celebrated and costly obsession with the Paige typesetter and his installation of the first private telephone in America in his own home are evidence of a lifelong enthusiasm for technological advancement. Overall, Twain's extensive knowledge of machines and their use is a practical knowledge, which is not as easy to document historically as theoretical knowledge and education but that nonetheless is a noteworthy accomplishment for an American writer.

At the time he wrote The Petrified Man, Twain had already accumulated an impressive amount of this practical education, everything but the inventing phase. He was also subject to ethnoscientific influences from the media and the public discourse around him. Two major influences among these are the Civil War and the publication of Darwin's *Origin of Species*.

The Civil War

The Civil War did not have the effect that people often assume it did—boosting scientific prestige, progress, and funding through weapons research. In fact, argues Robert V. Bruce in *The Launching of Modern American Science*, the Civil War was a huge setback for scientific research. The weapons that debuted in the war, such as the Gatling gun, were previous inventions, and no innovative war technology was developed in the early 1860s.[34] Instead, Southern scientific research was set back for decades with the destruction of facilities and the leveling of the economy. Even Northern universities could not get their funding back up to normal levels and begin to move forward with the organization of new programs until the mid-1870s.[35] Scientifically and technologically speaking, then, the war was a bust.

Twain did a brief stint with the Marion Rangers volunteer corps before deserting and heading west with his brother. Personally, he did not engage in a single campaign, but he communicated his discomfort with his role as a soldier in the burlesque "A Private History of the Campaign That Failed," which includes the eerie portrait of an enemy whom Twain and some fellow volunteers shot in the dark; he turned out to be an unarmed, nonuniformed stranger merely riding past their outpost.[36] His experiences as a witness to modern technological warfare also shaped the gruesome "Sand Belt" chapter of *A Connecticut Yankee in King Arthur's Court*.

Twain was not alone in his developing awareness of the dangers of technology when bent to the purposes of American imperialism, which he strenuously opposed. H. Bruce Franklin documents the distribution of this fundamental unease throughout Twain's culture:

> As Twain wrote, American culture was generating a contradictory vision of the relations between industrial capitalism and modern warfare, one that exalted weapons technology as the path to peace and progress. American popular fiction was shaping the cult of the superweapon—an invincible product of American ingenuity that would defeat all the backward and evil forces of the planet, thereby ending war and bringing about a global Pax Americana.[37]

Twain's contemporary and fellow examiner of the impact of the machine on American society, Henry Adams, rendered both the awe and terror of the situation in "The Virgin and the Dynamo" chapter of *The Education of Henry Adams*. Adams sees in the dynamo the destruction of everything pure and beautiful about civilization, as figured by the medieval virgin. Adams's imagery is more violent than, but akin to, the annihilation of the mysterious beauty of the Mississippi by Twain's technological knowledge of it, or the cleaving of Huck Finn's idyllic raft by the bow of the steamboat in *Huckleberry Finn*.[38] Twain was obviously responding to a cultural nervousness about technology and its easy adaptation to war, and he was responding to it as early as his desertion from the Marion Rangers.

Darwinism

In addition to the fallout from the Civil War, another powerful scientific trope pervaded Twain's culture, and that was Darwin's theory of evolution. An editor for the *Galaxy*, for which Twain wrote, exclaimed upon the "universal drenching" of literature and journalism in America with Darwinian ideas during the 1860s and 1870s.[39] Although Twain was a fan of Darwin's later *Descent of Man*, there is no evidence that Twain read the *Origin*, but most Americans did not. Instead, they relied on what they heard about it and what they read in the newspapers, including satiric reports and cartoons. American Darwinism was, for the most part, filtered through Lamarckianism, which stressed an element of design in development that made evolution more compatible with traditional Protestant beliefs.[40] It was also filtered through a strong bias toward organicism and individualism inherited from the American romantics and transcendentalists. Cynthia Russett argues that "romantic philosophy appealed to a more congenial scientific concept, that of the organism, against a less congenial one, the machine."[41] Thus, the idea of evolution as an organizing principle for all of life—especially when directed by a benevolent God, as figured by American Lamarckians—seemed a safe haven compared to the inhuman mechanisms of a clockwork Enlightenment universe.

However, evolution insinuated its own sort of mechanics into American scientific and social thinking, the fierce law of survival of the fittest. This was Huxley's nature, "red in tooth and claw," laid out in his 1893 *Evolution and Ethics*, a favorite on Twain's bookshelf. Russett describes the discomfort of American moral thinkers with the vanquishing of Newtonian mechanism by Darwinian evolution, which seemed like an "iron maiden presiding over endless panoramas of anguish and extinction."[42] Within Twain's lifetime, American writers were already dramatizing the cruel results of capitalist appropriations of survival of the fittest, or Social Darwinism: Upton Sinclair's *Jungle* (1906) is one of the most notable fables exposing the inhuman aspects of Darwinism as social policy.

American Darwinism, in short, sets up a pattern of embracing the elegant story evolution told about the development of self-sufficient life, while simultaneously shuddering at the cold amorality of a universe that was not designed but just *happened* at an astounding rate of attrition. This pattern shows up clearly in Twain's thinking, whether he came to it himself, borrowed it from Darwin and Huxley, or simply used parts of it to reinforce his own native cynicism about man's inhumanity to man. As mentioned above, Twain was sympathetic to the broad idea of evolution, as evidenced by his enthusiastic reporting of the conversations he had with his boarding-house philosopher friend McFarlane. His predisposition to believe in an evolutionary—and atheistic—model of life development, however, did not insulate the scientists who adopted Darwin's theory from his sharp, satirical pen. Twain satirized paleontologists at least three times—in "How the Animals of the Woods Sent Out a Scientific Expedition," The Petrified Man, and *3,000 Years among the Microbes*—for jumping to evolutionary conclusions on the basis of scanty fossil data.[43] Twain celebrated the idea of biological determinism, which reinforced his belief that people behaved no differently than animals, and perhaps worse. But at moments, the determinism of a Godless universe motivated merely by competition for resources seemed to knock the wind out of him.[44] Contrary to many scholars' beliefs that Twain simply succumbed to despair in a mechanistic universe, however, his hoaxing practices show Twain developing a coping strategy that asserts self-determination without necessarily buying into any established belief system—including evolution. His strategy is laughter, which remains both the simplest and the most complex response to the rhetoric of social determinism. The laughter also creates a bond between Twain and his readers, reasserting his, rather than science's, authority over them as their channel to the truth in the West. An examination of Twain's only scientific media hoax, The Petrified Man alongside his other media hoax, Empire City Massacre will show how these dynamics of self-determination and authority enter Twain's writings about science.

After studying The Petrified Man as a hoaxing event, I will use an OT-based approach to reconsider other, more traditional rhetorical analyses of Twain's hoaxing. And finally, I will develop a theory of Twain's use of laughter as a countermove against the rhetoric of social control, showing how this changes our perception of his writings about science in his later works.

The Petrified Man

Twain hired on at the *Territorial Enterprise* in the late summer of 1862 to fill in for the local editor, Dan De Quille (William Wright), who was visiting his family in Iowa. The first extant story Twain wrote for the *Enterprise* is a scientific hoax, The Petrified Man, which was printed in the *Enterprise* on 5 October 1862. However, due to a fire in 1875 that destroyed the *Enterprise's* archives, the only remaining copies of the story are twelve reprints in other area papers, beginning with the San Francisco *Evening Bulletin* on 15 October 1862.

The hoax can be examined in its entirety, since it is relatively brief. This is the text of the first reprint of the hoax in the *Evening Bulletin* on 15 October 1862:

> A petrified man was found some time ago in the mountains south of Gravelly Ford. Every limb and feature of the stony mummy was perfect, not even excepting the left leg, which has evidently been a wooden one during the lifetime of the owner—which lifetime, by the way, came to a close about a century ago, in the opinion of a *savan* who has examined the defunct. The body was in a sitting posture and leaning against a huge mass of croppings; the attitude was pensive, the right thumb resting against the side of the nose; the left thumb partially supported the chin, the forefinger pressing the inner corner of the left eye and drawing it partly open; the right eye was closed and the fingers of the right hand spread apart. This strange freak of nature created a profound sensation in the vicinity, and our informant states that, by request, Justice Sewell or Sowell of Humboldt City at once proceeded to the spot and held an inquest on the body. The verdict of the jury was that "deceased came to his death from protracted exposure," etc. The people of the neighborhood volunteered to bury the poor unfortunate, and were even anxious to do so; but it was discovered, when they attempted to remove him, that the water which had dropped upon him for ages from the crag above, had coursed down his back and deposited a limestone sediment under him which had glued him to the bed rock upon which he sat, as with a cement of adamant, and Judge S. refused to allow the charitable citizens to blast him from his position. The opinion expressed by his Honor that such a course would be little less than sacrilege, was eminently just and proper. Everybody goes to see the stone man, as many as 300 persons having visited the hardened creature during the past five or six weeks.

Reader expectations can be recuperated from this hoax by the same methods I applied to Poe's first hoaxing attempt. First, I will examine contemporary reactions to the hoax in newspapers and memoirs. These will be compared to Twain's writings about the hoax, what he claimed to have been trying to accomplish with it, and how he explained the construction of its rhetoric. Next, I will reconstruct the popular science article of 1865, since changes are bound to have accrued to the genre since 1835, and this portrait will reveal the generic expectations newsreaders might have had when coming to Twain's hoax. Finally, based on these collected expectations, I will make changes to the filter of expectations as constructed so far to account for historical change.

Contemporary Reaction to The Petrified Man

When Twain wrote in the "Memoranda" that "everyone was receiving [the petrified man] in innocent and good faith," he was exaggerating. Even if they ignored the obvious nose thumbing of the unfortunate petrifactee, several of the papers who reprinted it did so tongue-in-cheek. In *Early Tales and Sketches*, Branch and colleagues note that a majority of the twelve reprints were introduced in a straight-faced fashion.[45] Of the five reprints I was able to locate in the Berkeley microfilms of western papers, three were introduced ironically. I have indicated topoi associated with reading expectations by bolding them and have underlined conflict markers:

- A WASHOE JOKE.—The *Territorial Enterprise* has a joke of a "petrified man" having been **found** on the plains, which the **interior journals** seem to be copying in good faith. Our **authority** gravely says: . . . (San Francisco *Evening Bulletin*)
- THAT PIECE OF PETRIFIED HUMANITY.—The *Enterprise*, **published in Virginia City**, has the following, probably a hoax, about the discovery of a petrified man near Gravelly Ford, in Nevada Territory. It says. . . . (Sacramento *Bee*, 16 October 1862)
- A PETRIFIED MAN IN NEVADA TERRITORY.—The **Virginia City *Enterprise*** gets off the following sell about the **discovery** of a petrified man near Gravelly Ford, in Nevada Territory. (San Francisco *Alta California*, 15 October 1862)

The editors clearly rank highly the reputation of the medium—with a definite negative valence—in their decision to discount Twain's tale. So medium wins out over the novelty (indexed by words such as "found" and "discovery") of a petrified human being for these editors. It is safe to assume that they were not skimming the article, as they picked up the exposing details of the mummy's posture and recognized the story for a hoax. Internal coherence would therefore be highly ranked for these editorial readers, too, because the impossibility of a man being petrified while thumbing his nose is what gives the story away. So, the reconstruction of the editors' rejection of The Petrified Man, based on their commentary appears in table 4.1.

Table 4.1 Decision of Rival Editors about The Petrified Man

	Medium	Internal Coherence	Novelty
True	*	*!	
✓False			*

Even though deciding that the Petrified Man is a "false" science report requires the editors to deny that novel scientific reports are usually true, they are more concerned with the reputation of the *Enterprise* and the internal consistency of the story. The "!" by the violation for internal coherence, indicating that it is the fatal violation in the contest between the two interpretations, is a little misleading, for in fact, there is not enough data to determine which was the absolute deciding factor for the editors—the internal inconsistencies of The Petrified Man or simply the story's appearance in a local-yokel "interior journal." The vertical dotted line between the expectations indicates this indeterminacy. The convention in optimality theory is simply to indicate the "fatal violation" as far right as possible in a given level of rank, thus marking the first violation (*) that tips the scales and eliminates a candidate interpretation from the game.

In addition to the editors' own decision process, two other groups of readers and their reading expectations are invoked in the *Bulletin*'s comments: the *Bulletin*'s readers and the "interior journals'" editors and readers. These models and their implications for our common reading filter will be considered in turn.

The Bulletin's Projected Readers: Progression versus Comprehension

The editorial accompanying the *Bulletin*'s reprint of The Petrified Man projects a model of readers' expectations, similar to Poe's model of antebellum newsreader expectations in his analysis of the success of Locke's Moon Hoax versus Hans Phaall. Simply put, while the San Francisco editors believed their readers to share their parochial disdain for the shoddy journalism of the *Enterprise*, they feared that their readers' excited desire to progress through all the pertinent points of the story would cause them to read too quickly and miss the exposing details. We know this because the editors of the *Evening Bulletin* interpolated a bracketed exclamation mark "[!]" right after the description of the mummy's posture when they reprinted The Petrified Man so that readers who had been skimming would stop at the mark, go back, and reevaluate the position of the petrified man to see that he was thumbing his nose at them.

"Skimming" as a strategy to optimize expenditure of time and effort during reading is well documented in recent studies of scientific readers by Charles Bazerman and Davida Charney.[46] In addition, skimming has already been treated as an OT-type constraint-satisfaction process by Bertrand Gervais, a reader-oriented critic. He describes the process the *Bulletin* editors were trying to tamper with as a continual tension between two competing constraints, progression and comprehension. Progression expresses the desire of the reader to get through the material as quickly as possible. Comprehension expresses the reader's desire to understand the

Table 4.2 Editors' Projection of Their Readers' Interpretive Process

	Progression	Comprehension
✓True		*
False	*!	

details of what he has read. The reading activity proceeds as a negotiation between these two constraints, with reading generally proceeding as fast as will allow the reader to glean what he wishes to learn from the text.[47]

By signaling the exposing details in Twain's story, the *Bulletin* editors were educating their readers to read as they did, to rank their expectations the same. But the fact that the editors felt the need to educate their readers in this fashion indicates that the editors believed their readers read differently and too quickly. The editors' model of their readers' expectations is depicted in table 4.2. The editors hoped their warning sign would serve as a corrective to this reading habit, would force their readers to stop their headlong rush through the text and reread the details of Twain's story carefully enough to comprehend the paradox of the petrified man's posture.[48]

The "Interior Journals": Novelty over Medium

There is little doubt that San Francisco's media encouraged an image of the city as more savvy and cosmopolitan than the upstart mining camps in the Nevada territory. This attitude is apparent when the *Bulletin* editors distinguish themselves and their readers from the "interior journals" who were "copying it [the joke] in good faith." Their invocation of this contrast suggests that mining journals and their readers ranked their science newsreading expectations "backward" from the refined urban journals: in short, prizing novelty over the reputation of the medium and internal coherence. Judging if their projection was accurate will require a brief detour into the history of the *Enterprise* and an analysis of its relationship to its readership.

The *Territorial Enterprise* was "the largest paper in the West of the gold and silver rushes," according to Judith Yaross Lee ("Pseudo" 129), although the Great Fire of 1875 wiped out the newspaper's records, so circulation figures from that era are unavailable. It was the major paper in Virginia City and was read by miners, businessmen, politicians, and their families. It often courted its women readers with items specifically "for the ladies." The local miners relied on it for news of what was happening in other mines on the Comstock lode and for announcements of new claims and patents.[49] Politics was also big news, as Nevada's legislature was making a bid for statehood. Twain's brother Orion was the territorial secretary, and Twain himself covered the legislative sessions from 1862 through 1864.

However, the *Enterprise* was also known for having rather young bachelor editors who enjoyed drinking their "reporter's cobblers" and playing practical jokes on each other. Henry Nash Smith sums the situation up in this fashion: "Nevada journalism of the 1860's was nonchalant and uninhibited, and a report of the most commonplace event was likely to veer into fantasy or humorous diatribe."[50] Humorous stories, tall tales, and comic takes on local news filled in the gaps in the "real news" in every issue. Dan De Quille and Twain both used the paper to get in jabs at each other and at local politicians whom they did not like, as in the case of Twain using The Petrified Man to get back at Judge Sewall for some unrecorded offense. Twain was effectively run out of town on a rail because of a satirical article he wrote about the Carson City Sanitary Ball in 1864 that offended the ball's powerful organizers.[51]

Frank Luther Mott in the *History of American Journalism* argues that the editors of the *Enterprise* knew it was widely read and therefore used its substantial subscription as a platform to launch tall tales and hoaxes into the eastern media through the practice of clipping.[52] Certainly, from the evidence of the eastward migration of De Quille's hoaxes, which we will examine in the next chapter, this claim seems to be justified. How then, could readers of the *Enterprise* depend on its journalism at all?

Often the difference between the *Enterprise*'s readers "striking it rich" or not came down to whether or not they acted swiftly on new information, whether they received it through rumor or in the pages of the *Enterprise*; if they did not, someone else staked the big claim first. Paula Mitchell Marks found in her survey of miners' diaries from the gold rush era that the prospectors consistently appeared "ready to treat as a certainty the most tenuous of stories" in their desperation to make paydirt.[53] Granville Stuart, a contemporary historian of the gold rush, corroborates Marks' assessment in his own study of the Montana frontier: "Somebody would say that somebody said, that somebody had found a good thing and without further inquiry a hundred or more men would start out for the reported diggings."[54] Marks reprints a story from a miner's journal about a Montana town whipped into such a frenzy by these sorts of rumors that a man who was overheard in the saloon to say "I have got as good a thing as I want" was trailed out of town by twelve hundred citizens who thought he was leading them to paydirt. Imagine their embarrassment when they discovered, upon catching up with him at his cabin, that he had been referring to the Blackfoot woman he had just married.[55]

This panic for news was aided and abetted by an abnormal attachment to newspapers as lifelines to the civilized world from the remote mining camps. Miners waited impatiently for the delivery of the mails from the East containing papers that were in some cases already weeks old. Paula Mitchell Marks read in one Yukon miner's journal that he had once bought an outdated Seattle paper for fifty dollars (easily the equivalent of a week to two-week's gold finds for an average prospector) but then more than recouped his losses by charging his fellow camp mates a dollar each to hear him read the news from it.[56] In short, miners were a readership whose urgent need for novel information necessitated an unwarranted level of trust in Nevada journalism.[57]

Even more fortunate for Twain's hoaxing projects, the Jacksonian prospector was possessed by a tenacious attachment to promising rumors. He was obsessed with demonstrating that his sacrifice of his eastern life and family had not been in vain. James Tyson, a physician in a gold rush town, diagnosed his fellow miners as suffering from a singular stubbornness: "It is a remarkable and probably a commendable trait in the character of our go-ahead countrymen, to admit no statement contrary to their *preconceived opinions*, till by personal observation they have proved its truth or falsity" (emphasis mine).[58] Dan De Quille, who himself would go on to take advantage of this Quixotic single-mindedness among his fellow miners, reflected on his own problems with it one July Fourth as he rested among his fruitless diggings:

> As I lay on this knoll with the wind whirling the sand over me, and into my hair, ears, and eyes, sad thoughts of months and years spent in the chase of that Will-o-the-wisp, the 'big thing,' crowded themselves on my mind, and I was anything but 'gay and festive.' . . . I thought of the many long wild-goose-chases I had taken across deserts and over mountains. . . . How often had I sworn to be deluded no more by this deceitful phantom. But alas, for my resolution! . . . I fear that in spite of all the raillery . . . I will continue to follow the fiend till his designs against me are accomplished![59]

Even those miners who professed to disbelieve most rumors in the papers had a hard time keeping from getting caught up in the excited reaction to them. In his diaries prospector William Dennison Bickham astutely observed that the California papers' trumpeting of distant claims was largely at the behest of the transportation and provision companies, who profited immensely every time miners dropped everything at one claim site and raced each other to a new one.[60] However, after scoffing at a report in the San Francisco papers of a new claim at "Gold Bluffs," Bickham nonetheless feels a pang as a group of miners from his camp depart for the new diggings:

> This morning, whilst lying in bed, I heard loud shouts proceeding form the village and when I repaired thither I learned that they proceeded from a party of eleven men who had given

> the cheers as a parting salute to Murderer's Bar, they having started for the "Gold Bluffs." I hope they may have no cause to repent of their adventurous spirit, but reason bids me fear. *Nevertheless, if it were not for my having an interest here, I too should go* [my emphasis].[61]

As we can see, the *Bulletin* editors were not too far from the truth in their analysis of the miner readers of the western interior. Principally, these readers valued novelty highly because they needed fresh mining intelligence badly and were willing to overlook the spotty reputation of the mining newspapers to get it.

Summary of Reaction to The Petrified Man

The editorial comment on The Petrified Man confirms the ranking of medium over novelty that we established in the previous chapter. For these readers, the negative reputation of the *Enterprise* trumped their desires to "keep up with the times" and take novel scientific reports at face value. However, the comments also indicated that miners read science news for novelty and ignored the reputation of the medium; historical documents provided circumstantial evidence to support this projection of the editors. Finally, the editorial introduced two more expectations into the science newsreading process: progression and comprehension.

Twain himself was an acute observer of these reading patterns, and that sensibility appears to have governed his construction of his hoaxes. We will test the reader responses to The Petrified Man against his commentary on that hoax and on his other news hoax, Empire City Massacre.

Twain's Analysis of The Petrified Man and Empire City Massacre

Twain wrote in detail about his motivations and strategy in constructing The Petrified Man, and the best summary is his own:

> Now, to show how really hard it is to foist a moral or a truth upon an unsuspecting public through a burlesque without entirely and absurdly missing one's mark, I will here set down two experiences of my own in this thing. In the fall of 1862, in Nevada and California, the people got to running wild about extraordinary petrifactions and other natural marvels. One could scarcely pick up a paper without finding in it one or two glorified discoveries of this kind. The mania was becoming a little ridiculous. I was a brand-new local editor in Virginia City, and I felt called upon to destroy this growing evil: we all have

our benignant, fatherly moods at one time or another, I suppose. *I chose to kill the petrifaction mania with a delicate, a very delicate, satire. But maybe it was altogether too delicate, for nobody ever perceived the satire part of it at all.* I put my scheme in the shape of the discovery of a remarkable petrified man. I had had a temporary falling out with Mr. Sewall, the new coroner and justice of the peace of Humboldt, and I thought I might as well touch him up at the same time and make him ridiculous, and thus combine pleasure with business [my emphasis].[62]

Twain mentions many dynamics here that are significant for our study of hoaxing from the perspective of reader expectations. The key points, for our purposes, are as follows:

1. Twain was embarking on a project of social activism by trying to "foist a moral" on his readers.
2. He intended The Petrified Man to be a satire, not a hoax—the principal difference here, both in Twain's thinking and as discussed in chapter 1, being that in a satire the audience is "in" on the joke, whereas in a hoax the reader is a victim of the joke. Thus, he felt his satire failed because readers were duped by the story and missed the satiric bent.

Further, Twain recorded his account of readers' reactions to The Petrified Man. In the "Memoranda" article for the *Galaxy*, he makes strong claims about the dissemination of the hoax:

As a satire on the petrifaction mania, or anything else, my Petrified Man was a disheartening failure; for everybody received him in innocent good faith, and I was stunned to see the creature I had begotten to pull down the wonder-business with and bring derision upon it, calmly exalted to the grand chief place in the list of the genuine marvels our Nevada had produced. I was so disappointed at the curious miscarriage of my scheme that at first I was angry and did not like to think about it; but by and by, when the exchanges began to come in with the Petrified Man copied and guilelessly glorified, I began to feel a soothing secret satisfaction; and as my gentleman's field of travel broadened, and by the exchanges I saw that he steadily and implacably penetrated territory after territory, State after State, and land after land, till he swept the great globe and culminated in sublime and unimpeached legitimacy in the august,

> "London Lancet," my cup was full, and I said I was glad I had done it. It think that for about eleven months, as nearly as I can remember, Mr. Sewall's daily mail contained along in the neighborhood of half a bushel of newspapers hailing from many climes with the Petrified Man in them, marked around with a prominent belt of ink. I sent them to him. I did it for spite, not for fun. He used to shovel them into his back yard and curse. And every day during all those months the miners, his constituents (for miners never quit joking a person when they get started), would call on him and ask if he could tell them where they could get hold of a paper with the Petrified Man in it. He could have accommodated a continent with them. I hated Sewall in those days, and these things pacified me and pleased me. I could not have gotten more real comfort out of him without killing him.[63]

In 1937 DeLancey Ferguson, in what was probably the first scholarly treatment of the hoax, did a survey of the London *Lancet* for three years following the publication of the Petrified Man and found no mention of it in the paper. Likewise, Ferguson finds no mention of it in the eastern magazines he surveyed.[64] Twain was probably exaggerating the dissemination of his hoax, although locally it created a stir, as we have seen already.

What is really interesting about this hoax, however, is Twain's claim that the piece was in fact a failed satire. His analysis of its "failure" provides insights into the expectations of his readers, as he constructs them, and Twain is certainly a member of his own readership in this instance, being a pioneer and placer miner in Nevada. Twain wrote first of all that he expected the inaccuracies of Petrified Man to reveal it: "From beginning to end the 'Petrified Man' squib was a string of roaring absurdities, albeit they were told with an unfair pretence of truth that even imposed upon me to some extent, and I was in some danger of believing in my own fraud. But I really had no desire to deceive anybody, and no expectation of doing it. I depended on the way the petrified man was sitting to explain to the public that he was a swindle."[65]

Twain goes on to point out that locals were provided with further clues, such as the "people of the neighborhood" in Gravelly Ford, which was in fact a five-day ride into stark wilderness populated only by "a few starving Indians, some grasshoppers, and four or five buzzards out of meat and too feeble to get away."[66]

However, Twain contradicts his own claims of innocence somewhat as he admits the "unfair pretence to truth" he employed. Elsewhere in the same article, Twain once again confesses that he worked at making the Petrified

Man sound like an authentic news article: "So I told, in patient, belief-compelling detail, all about the finding of a petrified man at Gravelly Ford."[67] And when he wrote to his brother Orion on 21 October 1862, he made the imposture of his "squib" sound quite deliberate:

> (Between us, now)—did you see that squib of mine headed "Petrified Man?" It is an unmitigated lie, made from whole cloth. I got it up to worry Sewall. Every day, I send him some California paper containing it; moreover, I am getting things so arranged that he will soon begin to receive letters from all parts of the country, purporting to come from scientific men, asking for further information concerning the wonderful stone man. If I had plenty of time, I would worry the life out of the poor cuss.[68]

There is no evidence that Twain ever "arranged" for corroboration from scientists. But the claim shows him prepared to provide outside support for the hoax, which would not have been required if the story were simply a satire that flopped. Further, Twain published a follow-up piece in the *Enterprise* in November with the testimony of individuals who had been to see the "stone mummy" on display.[69] So, there are several solid reasons to doubt Twain's protestations that he did not intend to create a hoax. A further reason to doubt him is that almost exactly a year later, he wrote another one.

While The Petrified Man was Twain's only scientific media hoax, it was not his only hoax. The Empire City Massacre ran in the *Enterprise* on 29 October 1863. It was a horrifying report of a multiple murder/suicide that Twain designed in order to criticize a shifty policy on the part of mining companies of misreporting stock values—a brand of sharp dealing akin to the recent Enron scandal. In the hoax Twain claims a man driven mad by his losses on the stock market kills his entire family, cuts his own throat, and then rides into town and collapses in front of a saloon full of people, brandishing the scalp of one of his children. The hoax was apparently believed locally and resulted in a media furor once word leaked that Twain had faked it. According to Dan De Quille, Twain's fellow writer and roommate at the time, Nevada news editors called the story a "cruel and idiotic hoax," and California news editors threatened never to reprint another *Enterprise* item if J. T. Goodman did not fire Twain. Twain lost a great deal of sleep until De Quille reassured him that the whole thing would blow over eventually, which it did.[70]

Twain's own commentary on the hoax in his *Galaxy* piece is instructive for our purposes of reconstructing reader expectations because Twain was savvy to what is now called the "psychology of reading"—the cognitive process through which readers approach a text and interpret it. In that article Twain chalks up the phenomenon of his "satires" being read as hoaxes to three unanticipated factors: readers' extremely high valuation of sensation and novelty, the guiding principles by which his readers read science news, and the structure of that newsreading activity. All these observations corroborate contemporary reports of reactions to his hoaxes and anticipate current findings in reading psychology.

High Ranking of Sensation and Novelty

Twain would have agreed with the editors of the San Francisco *Evening Bulletin* that his readers put a lot of stock in novelty; to that assessment, Twain would have added sensation as well. Twain commented extensively on the power of the "wonder-business" of science to overwhelm readers' critical faculties. He claimed that this phenomenon, equivalent to a high valuation of the novelty and sensation, caused the "moral" of his satires to be missed. In the "Memoranda" he reflects, "[W]e never read the dull explanatory surroundings of marvellously exciting things when we have no occasion to suppose that some irresponsible scribbler is trying to defraud us; we skip all that, and hasten to revel in the blood-curdling particulars and be happy."[71] As to readers overlooking the clues he left in his "satires" and reading them as hoaxes, Twain cautions, "One can deliver a satire with telling force through the insidious medium of a travesty, if he is careful not to overwhelm the satire with the extraneous interest of the travesty, and so bury it from the readers' sight and leave him a joked and defrauded victim, when the honest intent was to add to either his knowledge or his wisdom."[72] Although it would be hasty to attribute any sort of "honest intent" to Twain's "satires," he makes it clear that the powerful sensational and novel aspects of a petrified man and a bloody massacre leave vivid images in readers' minds that are difficult to replace with subtle arguments about an overreliance on Darwinian paleontology or underhanded dividend cooking. About the Empire City Massacre, Twain laments, "To drop in with a poor little moral at the fag-end of such a gorgeous massacre, was to follow the expiring sun with a candle and hope to attract the world's attention to it."[73] Indeed, the "moral" chastising the dividend-cooking companies appears only in the last six sentences of the lengthy piece after some relatively dry biographical details about the murderer.

That "expiring sun" of novelty and sensation was Twain's lifelong whipping boy, according to Pascal Covici in *Mark Twain's Humor*. In that study Covici explicates Twain's "use of the hoax to ridicule the reader's penchant for collecting thrills."[74] He claims that Twain consistently used forms like the hoax and the tall tale as deflating pins for the balloonlike sensations of the "wonder-business" of popular science and any other mania that grabbed readers' attention through the media. Covici argues that

Twain felt Americans' slavish love of sensation was blinding them to the very real and unglamorous problems of poverty, racial inequality, and exploitation of ethnic minorities during Reconstruction.[75] This social function of the hoax will be discussed in the final section of this chapter with respect to Twain's activism.

Process of Reading

Twain identified another major cause for the failure of his "satires." He claimed the exposing details he buried in his "satires" did not surface into his readers' consciousness because his readers were reading too fast and simply missed them. His judgments echo those of the *Bulletin* editors reprinting his Petrified Man hoax: readers' progression expectation was overwhelming their comprehension of hidden details in the story. However, his statements were not mere armchair sociology; Twain actually conducted a sort of experiment in reading psychology after his second hoax.

In his discussion of Empire City Massacre, Twain reports watching a couple of farmers reading his article in an early read-aloud protocol experiment, just as Poe had spied on readers reacting to the Balloon-Hoax. Twain writes, "I saw that the heedless son of a hay-mow was skipping with all his might, in order to get to the bloody details as quickly as possible; and so he was missing the guide-boards I had set up to warn him that the whole thing was a fraud."[76] Twain goes on to mention that he had experienced this phenomenon before with an "agricultural" satire he wrote that was taken as the real thing. He theorizes, "Shall I tell you the real reason why I have unintentionally succeeded in fooling so many people? It is because some of them only read a little of the squib I wrote and jumped to the conclusion that it was serious, and the rest did not read it at all, but heard of my agricultural venture at second-hand."[77]

As we have witnessed with Poe's hoaxes, a hoaxer, unlike a satirist, relies on just this sensation-driven reading habit so she can secure most readers' confidence in the hoax while leaving behind little built-in clues that can then be pointed to later during the *elenchus* phase of the hoax. The nice fit of this hoax-specific strategy with Twain's lament about readers' carelessness seems further indication that he was crafting hoaxes when he wrote The Petrified Man and Empire City Massacre. His tears over his failed "satires" thus appear increasingly crocodilian.

Half of Twain's observation about his readers' reading styles attributes the failure of his "satire" to readers skimming too quickly to catch his "guide-boards." Naturally, the other half of his observation has to do with what readers were looking for in their headlong rush, what they were skimming over the rest of the text in order to reach. This has to do with industrial modes of reading news.

Newsreading

Twain also noticed that people consistently skipped to certain parts of his stories. As he pointed out with the Empire City Massacre, his readers were skimming to find the "bloody details" and the "blood-curdling particulars." More recent studies of newsreaders have corroborated Twain's observations. Researchers have found that readers indeed do not read news linearly but instead skim for information based on their desire to maximize novelty—things they did not know before. They are enabled in this procedure by the relatively stable structure of the news story: summary (including headline), main event, details, background, consequences, and finally comments.[78]

The form of the modern news article differs slightly, of course, from the popsci. criteria that we have observed in action. The modern article lacks the degree of "mystery" in its opening that the 1835 science news articles contained; the earlier articles also put background before the details of the main event in the form of failed attempts at discovery or understanding in the past; finally, the commentary in the early science articles often contains speculations that modern articles eschew in the interest of preserving an objective stance. However, as we will see in this chapter, the genre of the science news article was changing in Twain's time to conform more closely to the pattern Van Dijk observed.

If the form of the news article has changed over the last 150 years, however, the principle of the reader learning the form and then using it to adapt the reading experience to his goals remains the same. As mentioned in chapter 3, Rolf Zwaan has found that labeling the same text either a "story" or a "news article" for different sets of readers changes the way these groups read. His results showed that readers who believed they were reading news read for the details—the "who, what, when, where, why"—and had better recall for these details, and worse recall for fine details of language and presentation, as compared to readers who thought they were reading a fiction story.

Twain recognized that this structured reading activity was important for perpetrating a hoax. He noted that leaving his revelatory clues or "moral" until the end of the article caused it to fail as a satire because "the reader, not knowing that it is the key of the whole thing and the only important paragraph in the article, tranquilly turns up his nose and leaves it unread."[79] Twain evinces awareness of a front-loaded structure for news articles in his time. Since he had already had a similar problem with readers missing the "fine print" of the revelatory details of the Petrified Man, one would think he would have done something to foreground the revelatory details of Empire City Massacre if he genuinely intended to make a satire rather than a hoax. Dan De Quille claims in his memoirs to have suggested this to Twain during the composition of the story.[80] According to De Quille, Twain was defensive in the face of these suggestions, claiming that the small

Table 4.3 Skimming Guided By News Reading Conventions

	Sensation	Progression	News	Comprehension
✓TRUE				*
FALSE	*	*	*!	

inconsistencies he had sprinkled throughout the article—such as calling the murderer a "bachelor" even though he killed his wife and children, and mentioning a pine forest where locals should have known there was desert—were all "plain enough," and he refused to foreground the "moral" about dividend cooking that resided at the end of the story.[81] It seems clear that Twain intended to exploit what we now understand as a familiarity with the news article format in order to make Empire City Massacre seem like a "real" news story to his readers, in order to write a hoax, not a satire.

Since we have discussed Twain's farmers reading his Empire City Massacre over breakfast, we should consider how progression/comprehension and the familiar structure of newsreading might have interacted to produce belief in the hoax. The newsreading strategy operates on a similar level as progression and comprehension, since it determines which parts of the text are admitted to the interpretive process. Thus, we can argue for our newsreading strategy (news), progression, and sensation all driving the reading process at the expense of comprehension (of details), as depicted in table 4.3. An interpretation of the story as "true" wins out in spite of lack of comprehension of the text as a whole. This table thus represents Twain's belief about how progression/comprehension and news-based skimming interacted to make his readers jump to the wrong conclusion while reading his hoaxes.

Twain's Reputation as an Author (Authority)

Another peripheral issue to Twain's assessment of the reaction to the Petrified Man hoax is his readers' expectations about his behavior as an author at this juncture in his career. In fact his career was just getting started. The Petrified Man is the first confirmed piece of journalism that Twain wrote as a staff newspaper reporter. We know he wrote items for the local column around this time, many of them sarcastic. However, his legislative reporting was exact and trustworthy, if also occasionally critical and humorous.[82]

In February 1863, just four months after the publication of The Petrified Man, Twain began using the pseudonym he would use for the rest of his career. After that his readers had a heuristic for helping them decide if authority, or the reputation of the author, would have a positive or negative valence in their decisions about the truth of items in the *Enterprise*. Twain signed his serious political pieces "Sam Clemens" and his humorous bits "Mark Twain."[83] But plenty of his articles were unsigned, such as The Petrified Man, leaving the burden of deciding the values of both medium and authority on the reader.

Summary of Twain's Portrait of His Readers' Expectations

Twain, like Poe, seemed to think in terms of what readers expect or anticipate when he analyzed what went "wrong" with his "satires." Therefore, he makes the following significant contributions to our understanding of reader expectations of science journalism as practiced in the West:

- Novelty and sensation, representing the "wonder-business" of popular science, are perceived by both Twain and the *Bulletin* editors as the highest-ranked reader expectations of science news, thereby making them salient targets for his satirical attacks.
- The needs of core readers (miners) for novel information may corroborate a ranking of novelty above medium for miners who knew the *Enterprise* lied but who could not afford to pass up any tip on a new prospect.
- In signed pieces by Mark Twain/Sam Clemens, authority ("The reputation of the author holds") develops a schizoid status, as each name is associated with a different style of reporting: humorous/lying for Twain, and straight shooting for Clemens. Of course, many pieces, such as The Petrified Man, were unsigned altogether.
- The competing constraints of progression and comprehension govern the hoax-reading experience and may work in the hoaxer's favor due to readers' skimming habits as noted by both Twain and the editors responding to The Petrified Man. The competition between progression and comprehension may for some readers indicate a high ranking for sensation and novelty, but a low ranking or even a deactivation of detail and internal coherence, due to skipping these background details to get to the "juicy parts."
- A news-specific reading pattern may interact with the progression/comprehension dynamic to facilitate hoaxing and frustrate satire.

Twain's observations have been tested against the reactions of other readers of his hoaxes. Now, they will be compared to a profile of the popular science article in the 1860s. After this final step, the common reading filter of science newsreading expectations will be modified to account for changes in kairos since 1835 and for Twain's innovations in the genre.

The American Popular Science Genre in 1865

The Petrified Man adheres to the rhetorical form we have been studying in Poe's and Locke's hoaxes (the popsci. criteria) in several ways. The story includes painstaking detail in the description of the wooden leg and the attitude of the various limbs of the body and includes technical jargon such as "limestone sediment" (detail). There are sensational descriptions, including the chiming phrase "stony mummy." The story does display some humor, as in the ironic phrase "refused to allow the charitable citizens to blast him from his position." But this tone is not out of keeping with the *Enterprise*'s usual style of reporting actual events and discoveries in the Nevada territory in a joking manner, as discussed above.

We notice that the story is introduced without the typical "mystery" opening, though the mystery rhetoric is invoked later in the story with the phrase "strange freak of nature." Is this an early instance of the "who, what, when, where, why" rhetoric of the news article gradually phasing out the older, more elaborate style of the 1835 popular science article? It is of course impossible to determine this without a wide study, but a sample of eight contemporary journals, a count of the type of articles they contained, and a brief rhetorical analysis of the style of those articles may help suggest directions for future inquiry.

For this sample, as with the 1835 sample, I tried to select newspapers local to the Nevada and California region or eastern papers that De Quille and/or Twain were known to read. I have included the *American Journal of Science* and the *Scientific American* again for comparison with their earlier manifestations, to see what changes might have arisen the previous thirty years (an immediate observation is that both journals were publishing more science by volume in 1865 than in 1835). A few category labels have also

Table 4.4 1865 Media Surveyed with Number of Science Articles Per Issue

Magazine/Newspaper	Total Science Articles
Am. Journal of Science	150
Scientific American	117
SF *Daily Examiner*	26
NY *Sun*	17
NY *Times*	17
St. Louis *Missouri Reporter*	8
Sacramento *Transcript*	5
Virginia City, MT, *Post*	2

Table 4.5 Distribution of Categories of Science Articles across 1865 Media Sample

Major Category	Subcategory	Totals for Subcategories	Subcategory % of Total Science Articles	Major Category % of Total Science Articles
Pop. Sci.	Ad	24	25%	
	Blurb	1	1%	
	Joke	2	2%	
	Spectacle	5	5%	
	Almanac	4	4%	38%
Pop Tech.	How-to/Educ	2.4	3%	
	Blurb	3.9	4%	
	Ad	41.3	43%	
	Invention	3.7	5%	54%
"Pure" Sci.	Discovery	.2	.2%	
	Education	3.7	4%	
	Observation	.8	.8%	
	Review/Bio	2.6	3%	8%

changed. There were no metacommentary articles or poems in the popular science category in this survey, and I have added an almanac subcategory under popular science to reflect frequent weather, tide, and astronomical reports appearing in the daily papers midcentury. Under popular technology, I lumped educational and how-to articles together since most educational articles at this time were about how to build or do something rather than general histories. Tables 4.4 and 4.5 present the results of the survey. Because there is such a discrepancy in the circulation of the newspapers and journals considered, the totals by journal merely tell us that small town papers carried less science news than big city papers, an unremarkable result. However, the totals by category yield a snapshot of some changes in science journalism since 1835.

In 1865 popular technology was still by far the best-represented category in science news, and ads account for an even higher percentage of the total number of science articles included in each paper/journal. This could simply be because papers were carrying more ads in general as cities expanded. There was a 4.4 percent increase in announcements of spectacles or exhibitions since 1835, and this increase, while inconclusive, is consistent with the upsurge of Barnum and other medicine-show type entertainments since the 1830s. There was a marked drop-off in general educational items at the hard science level (10.8 percent), and this result could reflect the increasing professionalization of science media, including a general journal such as the *AJS*. This was accompanied by a decrease in blurb/factoid items in the popular science category (15.5 percent). Together, these results when put into the context of an overall 19% decrease in pure science news might reflect the pulling out of science from the popular press into more specialized journals, professional associations, and university courses. A more extensive survey would of course have to be conducted to confirm these results.

The rhetoric of scientific articles at this time is even more important to the project of updating our popsci. expectations to account for the thirty-year development in science journalism between 1835 and 1865. While the articles from the sample were still structured in a problem/solution/benefits format that provided a narrative of control over the awesome or uncontrollable, tables 4.6 through 4.8 demonstrate among other things that the "mystery" opening is not as sharply in evidence.

In short what we are seeing in this small sample is a greater conformation of science news to the rhetoric of regular news, with a structure very similar to that which Van Dijk found in his survey of news articles, discussed earlier: summary, story (main event, details of main event), then background, consequences, and commentary. The problem/solution schema particular to science news is still in effect, but the object now is not explaining wondrous and inexplicable forces of nature, but solving practical problems in health and industry.

Twain's The Petrified Man, at least as it was reprinted in the *Evening Bulletin*, conforms well to this new template. The petrified man is introduced straightforwardly, with "who, what, when, where" foregrounded in the first sentence. The main event is detailed next and then the "consequences/commentary" section is entered into with the words "This strange freak of nature has created a profound sensation in our vicinity." Empire City Massacre is structured much the same way. The popular science article was undergoing a change in Twain's time to conform to an event-oriented journalistic style.

These changes to the format of the popular science article will enter into the adjustment of the filter of newsreading expectations in the next section. Twain's considerable insight into the psychology of his readers offers us new perspectives on reader expectations in the 1860s. In addition, the reaction to The Petrified Man offers evidence of a multilayered news readership made up of groups of readers who each read the story differently to suit their differing agendas. All of this new historical evidence can be accommodated by the optimality-based model of reading expectations developed so far.

Table 4.6 Opening Structure of 1865 Popular Science Articles

Rhetorical Feature/Pattern	Example
Opening: Utility/the facts • Often gives a "who, what, when, where, why" snapshot in the first sentence • Often uses words such words as *success useful practical* • *wonder* language absent	**"Clams: How They Are Regarded and what is done with them by the Barn Island Club.** Clams are of various kinds, their usefulness is undoubted, their ameliorating effect upon human nature is undisputed; so say the members of the "Great Barn Island chowder Club," and so say we all. On Friday afternoon, in compliance with an elaborately elegant programme of invitation and arrangements, we went to Barn Island." *NY Times* SURGICAL FEAT.—Wednesday last, one of the most successful surgical operations was performed by our townsman, Dr. J. S. Glick, which speaks well for his skill. The particulars, we learn, are as follows . . ." Virginia City, Montana, *Post*

Table 4.7 Structure of "Problem" Phase of 1865 Popular Science Articles

Problem: sometimes still includes the "ignorance" function relating the scientific problem to a social or educational problem	"Perhaps the readers of the Times, who have enjoyed now and then a humble clam, would like to know how, on such a high and mighty occasion, those delicious bivalves are prepared." *NY Times*
	"Mr. Jas. W. Brown, in the summer of 1862, when in the employ of Ben Holladay received a gun shot in the left cheek, in a fight with Indians. . . . Mr. B's wound being dangerous, Ben Holladay took him to San Francisco to have the ball taken out, but the surgeons could not find it, supposing it to lay close to the occipital bone." *Post*

Table 4.8 Structure of "Solution" Phase of 1865 Popular Science Articles

Solution • sometimes still includes the "wisdom" function by naming cognoscenti • contains great attention to detail and occasionally still analogy	"First from a blazing fire the blaze was brushed, and the embers left bare. The hard clams by the bushel were put on, then soft clams . . . after which the entire mass is covered with a profusion of seaweed which keeps the steam in." *NY Times* "On Wednesday Dr. Glick casually observed that he thought he could find the ball. Mr. B. was ready immediately for the operation, which was successfully performed in twenty minutes. The Dr. first extracted a piece of the superior maxillary . . . and then found the ball to have lodged at the extreme lower portion of the ear and removed it by forceps through the cavity of the ear." *Post*
Benefit/Use: only practical benefits are enumerated, using language such as *successful perfect infallible*	". . . which nicely bakes the soft clams, nicely bakes the hard clams, cruelly stifles and beautifully colors the lobsters, and perfectly fits the potatoes for the dainty palate of the most epicuean Irishman." *NY Times* "We congratulate Dr. Glick on so well and successfully performed an operation." *Post*

Adjusting the Filter of Expectations to Account for Twain's hoaxing

After analyzing Poe's hoaxing, I developed the following working definition of a hoax: that it was a rhetorical exchange in a news medium between an author and a readership that served to sensitize readers to their collusion in the redefinition of reality according to ethnoscientific values. From the contemporary reaction to Poe's and Locke's hoaxes, I reconstructed the following provisional ranking of reader expectations about science news:

{Corroboration, Medium, Authority}>>
{Novelty, Sensation, Plausibility}>>
{Popsci., Foreign, Internal Coh.}

This notation represents three basic levels of ranking. Corroboration, medium, and authority were the strongest determiners of decisions about truth-value for newsreaders, equally ranked because these expectations did not compete with each other in the reaction data. Novelty, sensation, and plausibility represent readers' comparisons of the content of science news with the real world of their desires and experiences (we are dealing with a rough-grained plausibility here, like reader judgments about alchemy as a plausible field of scientific innovation); these expectations formed a mid-strength filter on truth-value judgments. The weakest constraints on truth-value decisions when reading science news (or science hoaxes) were textual and generic expectations: popsci., foreign, and internal coherence.

After analyzing the response to The Petrified Man and the Empire City Massacre, both in Twain's commentary and in the reprinting editors' introductions, important modifications need to be made both to the definition of hoaxing and to the ranking of reader expectations to reflect the elapsed thirty years in hoaxing practices. Twain offers us a valuable vantage point into hoaxing via his analysis of why his "satires" failed and turned into media hoaxes. First and foremost, his denial of deliberate intentions to hoax his readers makes us further revise our notions of author intentionality in hoaxing.

After Poe's cagey self-construction as a hoaxer in the case of M. Valdemar we recognized that initial authorial intentions interact with reader expectations of medium and reader responses over time in constructing a hoax. An author has many opportunities during a hoaxing exchange to claim responsibility for the hoax. In the case of Twain's denials of having deliberately crafted a hoax, we saw that textual and contextual evidence can argue against an author's claims of innocence. The evidence in his letter to his brother of Twain's gleeful plotting to further the effects of the hoax on Sewall, coupled with his followup of the The Petrified Man with further hoax material both argue that Twain was engaged in more than merely post hoc play with the unexpectedly credulous reaction to his "satire." Further, Twain went on to write the Empire City Massacre without correcting any of his rhetorical "failures" in The Petrified Man. He also wrote, during this time and after, many successful satires that no one took for hoaxes. It seems probable that Twain had intended to catch readers off-guard with The Petrified Man all along and humiliate them for their naïve trust in paleontologists and geologists. A satire, after all, as we saw in the first chapter, pits both author and reader together against a socially superior target that needs a comeuppance. A hoax, however, is "the sort of scientific humor that aims directly at the audience's ignorance" and reveals it to them.[85] The Petrified Man was called a hoax publicly by several of the reprinting editors, and Twain himself revealed it to be such in his letter to Orion and in the "Memoranda" article for the *Galaxy* in 1870.

However, Twain's claim that he "accidentally" produced a hoax is crucial to our study of author intentionality in hoaxing because it reveals once more that a hoax is not a monologic statement by an author but rather a rhetorical interaction, residing only in the coordinated activity among author, medium, and readership. Twain's point is that even if he did not intend a hoax, his audience's belief in his story created a hoax anyway. While the author's intention can direct readers' experience to a certain degree, it still does not constitute the hoax.

What Twain has tuned us into is the realization that reader desires and expectations are powerful determiners of hoaxing events. The *Enterprise*'s core readers needed new information more than they needed to save face. This led them to place trust in reports that the paper's reputation clearly did not warrant. This is not merely a repetition of the old saw, "People believe what they want to," but an opportunity to witness how specific readerly expectations, especially sensation and novelty and medium and authority, can be ranked differently based on different reader desires. We saw, through Twain's perspective and the perspective of rival editors, how different groups of *Enterprise* readers might be characterized by their different ranking of these expectations: miners would rank novelty over medium; most of the other core readers would simply suspend medium in their judgments; and "outsider" readers such as competing editors and eastern readers would rank medium over novelty. Further, for all readers who ranked novelty highest, entertainment (reading hoaxes purely as entertainment) would be deactivated because of their need to decide upon and use the information presented in the newspapers. Editors and eastern readers were the only ones with the luxury to rank entertainment highly and suspend truth judgments about Twain's story.

Twain's final contribution to the redefinition of hoaxing is the awareness his commentary raised about the interaction of psychological constraints with interpretive expectations in the reading process. Constraints such as progression and comprehension do not directly participate in decisions about truth-value, but they do act as an early filter on the information that gets admitted to the decision process. In addition, a high-ranked desire for sensation may both interact with attention-allocating expectations while also assisting in post hoc judgments about truth-value. After an analysis of The Petrified Man and the commentary surrounding it, I propose a new, dual ranking of reader expectations. The second tier represents the interaction of sensation with the new psychological reading constraints:

{Corroboration, Authority}>>
{Novelty, Sensation, Plausibility}>>
{(Medium), Popsci., Internal Coh.}

{Sensation, News, Progression}>>
Comprehension

We will analyze the first tier of expectations first, taking them level by level in order to be clear about the changes they register between newsreading in Poe's era in New York City and newsreading in Virginia City in 1865:

1. First level: Corroboration and authority. Corroboration remains in its position for lack of evidence for demotion. The privileging of authority in this new ranking is also not a change from Poe's era; however, it reflects a slightly different dynamic between author and reader. Twain's reputation was an excellent predictor of truth or falsehood for his readers since he was fairly consistent with signing "Sam Clemens" only to his serious political reporting; anything signed "Mark Twain" was obviously of low truth-value. Of course, just because authority was the highest ranked does not deactivate the rest of the expectations. In the event that authority did not decide the issue conclusively, readers resorted to the midstrength expectations to help them decide about truth-value.
2. Second level: {Novelty, sensation, plausibility}. This midstrength level reflects the instincts of both Twain and his competing editors that their readers attached high truth-value to what was new or amazing; it also reflects the needs of miners for a constant stream of "inside" information. Plausibility remains in this echelon due to lack of evidence in reader reactions for upgrading or downgrading this expectation.
3. Third level: {(Medium), popsci., internal coherence}. These expectations reflect internal and external measurements of the story against itself and against previous experience with the science news genre in other words, the elements of "common sense" readers had to use to make truth-value judgments in spite of conflicting or absent cues from the medium and author—

> provided they had not already been swept away by the "extraneous interest" of the story, in Twain's words. The biggest change since Poe's era is the downgrading and partial deactivation of medium. In this ranking, medium is presented in parenthesis to reflect the *Enterprise*'s habitual mixing of fact and fiction that rendered its reputation an unreliable barometer of truth for most readers; medium remains provisionally in the ranking because miners did have to weigh the tempting new information the paper published against its reputation for lying (medium with a negative valence). Foreign has been deactivated due to the absence of European characters from Twain's hoaxes.

The second tier of rankings expresses the way in which newsreading strategies (news), the excitement of sensation, and the desire to get to the good stuff (progression) often defeated comprehension of textual details, thus actually facilitating the hoax's initial bid for credence. For this reason, another of the "morals" that Twain embedded in his hoaxes, aside from "don't believe paleontologists" and "don't get swept up in poor investing schemes," could be stated "always read the fine print."

Applying the Analysis to Problems in Twain Scholarship

As with Poe, several Twain scholars have researched his hoaxing and attempted to define it and examine possible influences it may have had on other genres of his work. These analyses generally break down into two categories: some extend an analysis of Twain's media hoaxes, like The Petrified Man, and Empire City Massacre to find hoaxing behavior in Twain's fiction; others use the trope of illusion and humiliating revelation in the hoaxing activity to psychoanalyze Twain's relationship to science and technology, especially in the final years of his career. My approach—clarifying the relationships with science and readers that Twain constructs in the hoaxes first, and then tracking the development of these relationships through his later "scientific" fiction—validates the rhetorical methods of some of my predecessors while revising their tendencies to read Twain's late-life depression back onto his scientific rhetoric.

The scholarly tradition that reads the hoax as an organizing trope of Twain's other fictions can be represented by Joan Belcourt Ross, Lawrence Berkove, and Pascal Covici. Their project, overall, is hampered by a loose definition of hoaxing that fails to distinguish Twain's literary projects from his social projects. The necessity of covering the "several hundred" examples of hoaxes that Joan Ross finds in the Twain oeuvre leads her to create a vague definition of the practice that cannot help but contradict itself when applied to moments as diverse as Huck playing a practical joke on Jim in *Huckleberry Finn* and the identity switch at the heart of *The Prince and the Pauper*. For example, Ross initially defines a hoax by saying it provides its writer with human comforts—such as security and money, but two pages later she claims that hoaxers risk the quite discomfiting results of "shame, humiliation, and in extreme cases, death" by perpetrating their hoaxes.[86]

Lawrence Berkove attempts to avoid self-contradiction by constraining his definition of hoaxing. He identifies *A Connecticut Yankee in King Arthur's Court* as a "tragic hoax" as opposed to a comedic hoax like Petrified Man. He writes, "Readers expect the target of the hoax to be one or more of the characters in a work of fiction; it always comes as a shock, when they learn that they are the author's real target."[87] The central problem with this definition is that once readers know they are reading a novel like *A Connecticut Yankee* and have therefore decoupled their interpretation of it from their construction of the real world they live in, it is impossible for the author to make them the "target"; therefore, it is impossible to hoax them in the same way that Twain hoaxed his farmer-readers or the editors who blithely reprinted The Petrified Man as news. Reality as readers know it—that is, the reality in which they pick up their kids from school, go to church, or buy stock—does not change to match the world of the science fiction novel they have just finished. The realities of those who believe media hoaxes, on the other hand, change until the hoax is revealed, and perhaps even after that.

Pascal Covici's conception of hoaxing is, of this school, the most congenial to the aims of this project; he thinks of hoaxes in terms of reader expectations. He was the one who first pointed out that Twain was using hoaxes such as The Petrified Man and Empire City Massacre as weapons against postbellum readers' tendency toward sensationalism or "collecting thrills." Covici concluded, "The hoax is aimed at the assumptions that make the romantic appreciation of sensationalism possible,"[88] because the hoax must match those assumptions, at least initially, in order to achieve its humbling effect.

However, Covici's definition of hoaxing is still too broad. He wishes to recuperate an excised section of *Life on the Mississippi* concerning a grisly balloon journey as a lost Twain hoax. When he points out that the story is almost certainly a parody of Poe's "M.S. Found in a Bottle" and Hans Phaall, he is in fact correct. With its exaggerated descriptions, like "dead people in all possible stages of greenness & mildew, & all of them grinning & staring . . . & a lot of dried animals of one sort & another,"[89] and its being marked for inclusion in a hyperbolic memoir, the tale is indeed a parody rather than a hoax; it cannot play with its readers' conceptions of reality. Neither can the other "missing hoax" Covici wishes to include in the Twain hoax canon, the parodic opening of Twain's "Double-Barreled Detective Story" where Twain sets the mood with "a solitary esophagus sle[eping] upon motionless wing."[90] In this second case, Covici argues that Twain's "double-speak" with scientific terminology like the word "esophagus" "suggests the formal prerequisites of any hoax,"[91] and he is again right. Choosing (or creating) words that are read over and accepted without comprehension or analysis by readers who just assume the concept is over their heads is perhaps the most fundamental move toward building a media hoax. But the "esophagus" paragraph and the excised balloon chapter of *Life on the Mississippi* are both literary parodies, appearing in literary contexts where they are disabled from altering readers' expectations of reality. So, Covici's criticism of Twain's hoaxing, while the most rhetorical and perceptive of the school we have just reviewed, still conflates playing with literary conventions and playing with epistemology. This mistake obscures the choice that Twain made to address his readers' construction of reality with his hoaxes and his social criticism, as we will see momentarily.

The second category of Twain hoax scholarship is the psychoanalytic category. The basic critical move made by the foundational scholars in this school is to use Twain's disappointments with scientists and machines to argue that he, though initially gulled by a vision of utopian progress through science, became disillusioned with science in his later years and feared it would be used by people merely to destroy each other. The more rhetorical angle of this view holds that Twain performed hoaxes to dramatize to himself and to his readers his sense of betrayal by promises of technological utopias that delivered only dystopic self-annihilation. This approach to Twain's relationship to science tends to get the cart before the horse, reading Twain's bitter losses and disappointments with science and medicine later in his life back onto his scientific rhetoric.

The psychological school of Twain's scientific philosophy was opened by Hyatt Waggoner, who claimed Twain came to the conclusion that self-determination was an illusion. "Mark Twain lived the last fifteen years of his life a bitter pessimist and a philosophical mechanist. . . . He came to think of man as a machine buffeted by an indifferent, if not hostile, mechanical universe."[92] This essential point has been repeated with small variations in studies of Twain's science by Tom Burnam, James Wilson, and Sherwood Cummings. Shelley Fisher Fishkin instead invokes "ambivalence," the prevailing trope in criticism of the American romantics' attitudes toward science, to label Twain's own scientific philosophy. She claims that Twain's disappointments with science, figured by the death of his brother Henry in a steamboat explosion and Twain's failed investment in the Paige typesetter, made him ambivalent toward science's intense promise for improvement of American life on the one hand and its "potential for dehumanization and devastation" on the other.[93]

Joan Belcourt Ross and Lawrence Berkove extended this mode of criticism in the last phases of their more rhetorical analyses of Twain's hoaxing. Both of them take the hoax as the organizing principle of Twain's scientific philosophy and cosmology. Berkove figures the "Sand Belt" chapter of *A Connecticut Yankee*, in which Hank Morgan ironically ends up barbecuing eleven thousand knights on his electric fences after working for years to "rescue" feudal England from its barbaric ignorance, as a dramatization of Twain's own belief that humans are merely the "butt of God's rather grim practical joke or 'hoax.'"[94] Ross's arguments about the power of the hoax in Twain's personal philosophy are not centered specifically on his view of science, but on his determinist view of human nature in general, the "propensity of all men, everywhere, consciously or unconsciously, to be implicated in the creation and/or the sustaining of illusion."[95] Ross claims that Twain used laughter to try to shock his readers out of this state into fleeting moments of self-realization, an interesting argument we will turn to in the final section. The consensus of these scholars' readings of the hoax in Twain's scientific and moral philosophy is that the hoax became the organizing trope of his thought—that Twain came to view human beings as gullible victims of a universal hoax and that, except for the brief respite offered by laughter, there was no way out of this grim cycle of illusion and disillusionment.

These theories tell us less about Twain's views on science than they do about our own desires as scholars to reduce Twain's personality, philosophy, and rhetorical practices to formulae. Unfortunately, this reduction accidentally pares away the social aspects of Twain's hoaxing. The hoax is an essentially social transaction, requiring for its dual effect of illusion/revelation all of the disparities in knowledge and imperfect transfers of information that obtain among author, medium, and audience in a newsreading culture. As a social transaction, the hoax was a conscious rhetorical choice of stance for Twain, not an inevitable side effect of his "mechanistic" world view. Twain chose many other genres of interaction with his readers as well: satire, novels, essays, editorials, travelogues, and lectures. Why not examine Twain's science through these rhetorical lenses instead? Scholars are attracted to the idea of viewing Twain's personal philosophy through the hoax because it presents an appearance/reality dichotomy as well as a metaphor for social control, and these issues definitely mirror fears that Twain harbored about the role of science in human society. However, these scholars have not yet succeeded in establishing any kind of necessary connection between Twain's use of the hoax and Twain's moral or scientific philosophy. In fact, the one scientific media hoax Twain actually pulled off, The Petrified Man, was written when Twain was young and optimistic. It does not discuss machines or the mechanistic world view that these scholars believe was Twain's last resort in his later years after he rejected the illusion of the scientific salvation of humanity.

The promising aspects of these approaches to Twain's philosophy of science, then, are relatively restricted. In the final phases of Joan Belcourt Ross's argument, she inquires into what sort of relationship Twain was attempting to build with his readership via the rhetoric of the hoax. That question will be the focus of the next and final section of this chapter. In establishing the significance of the hoax in Twain's scientific thought, we must work against the grain of Twain scholarship to this point. Instead of projecting Twain's mechanistic depression backward and reading it onto his scientific hoaxing behavior, we must work from his early hoaxing forward through his thinking, traveling along the threads of arguments introduced by the hoaxes to see how they are developed in Twain's writing about science.

Relationship of the Hoax to Twain's Scientific Thinking

We have already discussed Twain's early hoaxes, The Petrified Man, and Empire City Massacre, in some detail. Identifying the arguments Twain makes about science here is relatively easy, as Twain's philosophy of science is simple and relatively undeveloped at this stage. The Petrified Man argues that both scientists and lay readers jump to conclusions, that the desire to prove a theory (Darwinism) or the desire for titillation makes people overlook facts that would lead them to a different and more sobering conclusion if they took the time to consider them. The Petrified Man also performs two additional arguments about science in culture: one, that scientists are competing with journalists for the authority to create the West for readers; two, that the media both exacerbates this problem by reprinting sensational science news willy-nilly and provides a potential remedy in the form of gate-keeping editors who exercise their common sense. However, Twain's hoax dramatizes the extreme difficulty of using rhetoric to sensitize readers to their naïve assumptions. Namely, if they will not slow down long enough to read what has been written, the best and most telling argument against their naïveté can turn easily into a confirmation of it. A brief review of three of Twain's important later works concerning science, *3,000 Years among the Microbes*, *Connecticut Yankee*, and *The American Claimant*, will reveal how these early fibers of scientific rhetoric are worked by an older and wiser Twain.

3,000 Years among the Microbes

This unfinished novel promotes a fractal model of human society through the investigation of one of its microcosms: germ society. Twain's basic postulate is that every being is made up of a society of smaller beings who are not aware that the "universe" they inhabit is really just a larger, more complex organism that, in turn, belongs to a society of its own. The narrator is a cholera germ who calls himself "Huck," among other names, and he alone knows that all of his germ compatriots in all their levels and castes of society are merely cells in the body of a diseased tramp named Blitzkowski; Huck bears this privileged awareness because he was a human, and a scientist, before he was accidentally turned into a germ by an "alchemist." The implication pervading Huck's stories about his life among the germs is that humans themselves are various types of germs—good and bad—in a greater organism they call "the universe."

Twain's research into microbiology, manifest in this book, is impressive for a writer in the early twentieth century. His ecological sensitivity is noteworthy, too, as he considers the ignorance of disease germs as they wreck their "environment." A yellow-fever germ friend of Huck complains that the tiny microorganisms that infest the germs themselves are not aware of the pain they are causing their hosts; if they knew, they would stop. Huck muses at this, "You notice that? He did not suspect that he, also, was engaged in gnawing, torturing, defiling, rotting, and murdering a fellow-creature—he and all the swarming billions of his race. None of them suspects it."[96] Huck goes on to make it clear that this statement applies to human ecology as well:

> It hints at the possibility that the procession of known and listed devourers and persecutors is not complete. It suggests the possibility, and substantially the certainty, that man is himself a microbe, and his globe a blood-corpuscle drifting with its shining brethren of the Milky Way down a vein of the Master and Maker of all things, Whose body, mayhap,—glimpsed partwise from the earth by night, and receding and lost to view in the measureless remotenesses of Space—is what men name the Universe.[97]

Huck persists with his human scientific activities even as a germ. He and some friends put together a scientific society on the basis of having "excavated" a calcified flea imbedded in one of Blitzkowski's arteries. They gather at the "fossil mine" and pick over the bones as human paleontologists would with a dinosaur. Here, it appears Twain intended a second satire of paleontologists who build monsters out of a single bone chip, similar to the mocking portrait he crafted in "How the Animals of the Woods Sent Out a Scientific Expedition," but he cut all eighteen pages of it from the manuscript. He does not excise, however, a stinging satire of scientists for being so absorbed in their methods that they miss the truth. Huck explains to his scientist germ friends that he was once a human and that the organism they inhabit is part of a whole other and bigger universe. He is either scoffed at for a liar or lauded for his beautiful "poetry."[98] Twain also satirizes the mercenary turn of science as its shoulder inevitably bends to the wheel of capitalism in America. Huck discovers a gold mine in one of Blitzkowski's molars—actually, he imagines it is there and convinces all the other scientists that it is, too. But the more he considers working the mine with them, the smaller and smaller a share of the profits he is willing to give his friends.[99] The manuscript ends there, leaving a perhaps unfair but powerful impression that Twain believes the ultimate goal of American science is not finding the truth but making a buck. This rather cynical ending leads Beverly Hume to speculate that *3,000 Years among the Microbes* is an exercise for Twain in exorcising his own get-rich-quick demons in the wake of the Paige typesetter disaster.[100]

A Connecticut Yankee in King Arthur's Court

This novel is probably the most discussed in the reconstruction of Twain's views of the role of science and technology in human society. It is the cornerstone of psychoanalytic arguments that Twain felt hoaxed by technology and by the universe, particularly the final chapter, "The Sand Belt." For a book that is in general a humorous satire on systems of primogeniture and aristocracy, which Twain found residually operative in Southern plantation culture, the straight-faced brutality of the final chapter is a shock. But it is also an indication that the "improvements" that have come before—Morgan's caste-leveling schools and factories, his capitalism, his introduction of Victorian hygiene and work ethic—are in the end deadly improvements, or at the least null improvements, as Morgan destroys any trace of them that history might have hoped to find.

H. Bruce Franklin points out that this sharp denunciation of the dystopic future offered by science is not the only social criticism of science Twain worked into the "Sand Belt" chapter of *Connecticut Yankee*. Franklin argues that Twain, innovating the genre of time-travel fiction with Morgan's trip back to feudal England, institutes a new social time scale, one based on technological advancement rather than chronology. It is a time scale Twain senses underlying the myth of progress in his own culture, and in *Connecticut Yankee* he rejects its utopic teleology:

> Hank's apocalyptic weapons resolve the paradoxes of time travel by destroying everything that the nineteenth century has anachronistically introduced into the dark ages. But this resolution itself is paradoxical. The science and technology that mark progress, that distinguish forward from backward in time, become the means to annihilate all that humanity has created. Thus they display their potential to transform the future into the prehuman primeval past, that is, mindless oblivion.[101]

From this viewpoint *Connecticut Yankee* is not just a criticism of the damaging power of technology bent to the human desire to dominate others. It is also a *refutation* of the entire social epistemology resulting from equating scientific advancement with human development.

The American Claimant

This novel is another roundabout attack on aristocratic systems—this time in the guise of a young earl's son who, to resolve a century-long dispute between the currently recognized Earl of Ross and an American "claimant" to that title, gives up his inheritance to come to America and live as a common American. Technology figures in this book in a positive and humorous light in comparison to most of Twain's later works. Colonel Sellers, the American claimant, spends half his time writing letters to the Earl of Ross in England pleading his claim to the title; the other half of his time, he spends inventing. Both activities are pure comic relief in Twain's novel. The invention that Sellers is dead serious about—a system for reanimating corpses for the purposes of soldiery and slave labor—amounts only to black comedy. However, the items he tosses off in his spare time—like a little tangle puzzle called "Pigs in Clover"—ironically make him loads of money, which he merely squanders in the pursuit of his reanimation system.

In addition to comedy, invention serves as a serious symbol of the positive face of American democracy in the novel. The young Tracy (the American name the young British Ross adopts after all evidence of his true identity is destroyed in a hotel fire) visits a Mechanics Club debate in Virginia and is awed and pleased at the democratic construction of knowledge he finds underway there. In his journal Tracy reproduces the speech given by a club member on the value of mechanics as citizens, over and against college-educated scientists:

> It can no doubt be easily shown that the colleges have contributed the intellectual part of this progress, and that that part is vast; but that the material progress has been immeasurably vaster I think you will concede. Now I have been looking over a list of inventors—the creators of this amazing material development—and I find that they were not college-bred men. . . . It is not overstatement to say that the imagination-stunning material development of this century, the only century worth living in since time itself was invented, is the creation of men not college-bred. We think we see what these inventors have done; no, we see only the visible vast frontage of their work; behind it is their far vaster work, and it is invisible to the careless glance. They

> have reconstructed this nation—made it over, that is—and, metaphorically speaking, have multiplied its numbers almost beyond the power of figures to express.[102]

The speaker proceeds to a calculation of how the machines that inventors have created have multiplied manpower. "You look around you and you see a nation of sixty millions—apparently; but secreted in their hands and brains, and invisible to your eyes, is the true population of this Republic, and it numbers forty billions! It is the stupendous creation of those humble, unlettered, un-college-bred inventors—all honor to their name."[103]

In light of Twain's background as a journeyman printer, riverboat pilot, and novice miner, it is not surprising to find his young British protagonist celebrating the advantage of tradesmen over the college educated. However, interestingly, Tracy also enthusiastically records the speaker's praise of the American media for its work exposing the schemes and illusions of the elite classes in America: "For its mission . . . is to stand guard over a nation's liberties, not its humbugs and shams."[104] This equation of the press with the revelation of "shams" perpetrated by powerful classes in America is crucial for our discussion of Twain's social project of hoaxing.

Analysis

In Twain's later works we find confirmation and development of nearly every strand of argument about science present in The Petrified Man with some additional lines of argument added. Twain still believes scientists are hurting themselves as much as the public when they make simple judgments about complicated systems motivated more by their commercial and political agendas then by an authentic hunger for truth. He still believes the media can help expose scientific and other social humbugs; however, he still respects and fears the power of readers/hearers to extract uncontrollably many interpretations from the same discourse, as illustrated by the "beautiful palace/beautiful lie" conundrum of Huck's exposition of the "real" human universe in *3,000 Years among the Microbes*.

To these lines of argument Twain has added others based on his life experience. They include an enthusiasm for invention in the spirit of the self-made Jacksonian tradesman—evident both in *American Claimant* and *Connecticut Yankee*—counterbalanced by a very real horror of what American inventions can do when put to the task of "correcting" the backward ways of less technologically developed countries. This "ambivalence" is what has led psychoanalytic scholars to identify Twain's experience with science and technology as the last straw in confirming his late-life depression. However, there is another, more rhetorical way to construct Twain's relationship to science suggested by his own works.

Twain's treatment of science in "The Mysterious Stranger" manuscripts, written in his dark later years, tells a slightly different story about Twain's beliefs about the relationship between science and human society than has regularly been assumed by scholars. In this work Little Satan argues that humanity damns itself with its "moral sense" and simply uses the products of science to help.[105] It is human nature that is to blame for the inability of the human race to progress past the brutal dominance of those who think and live differently. Science and technology simply aid that project both philosophically. Some of the same scholars who see Twain as a bitter mechanist do acknowledge this turn in Twain's thinking. Waggoner writes that Twain "used science to reinforce his thinking."[106] And James Wilson concludes after reading the "Mysterious Stranger" that Twain believes the worst of "man": "Science here is . . . merely a handmaiden to his depraved nature, creating a world even worse than the one preceding it."[107]

But at this point, this line of argument reaches a disconnect for most Twain scholars, for they generally evince a belief that Twain was not a social activist—a social satirist, certainly, but not someone who dealt seriously with contemporary social issues and tried to persuade Americans to adopt certain solutions. According to this view, when Twain said that science would not destroy human society but that human nature would, he was certainly criticizing, but not constructively. Indeed, Twain's pessimistic view of human nature, his determinism and mechanism, would not seem to foster any kind of ideal of progress or improvement for the human lot. But reexamining the question of what kind of relationship Twain was building with his readers through his hoaxes, coupled with a more careful analysis of Twain's social activism, leaves a very different impression. Twain did indeed have a social agenda, "to help a man to see himself true."[108] This turns out to be not merely a static goal but also a constructive one, accomplished through the social mechanics of laughter. To arrive at this conclusion, we will return to Twain's early roots in frontier humor and its social function, review the modifications he made to that mode with his media hoaxes, and consider how this activity meshed with his social criticism in his later years.

The Social Mechanics of Laughter

Laughter is an extremely specialized reaction to an argument of difference, specifically, a difference between our assumptions and a revelation of how things "really" are or how they are "really" being constructed for us by others. Laughter, especially in America, is a response to juxtaposition of the reader/viewer's assumptions with a very different presentation of reality—

a big man riding a bike that is much too small for him, a man wearing women's clothes, a baby talking with an adult's voice, a pack of wolves chasing a marching band instead of a herd of reindeer. In all of these cases, we perceive a gap between our a priori assumptions about the world and a surprise revelation of how it is really working at this moment (in irony, we find it works exactly counter to our assumptions). The perception of this gap triggers laughter as a response. For some of us, the laughter also expresses a desire to close the gap, to adjust our assumptions so as not to be caught off guard again. For others of us, the laughter is pure play, an appreciation of the imperfections and incompleteness of our understanding, a good-natured sympathy with the "joke being on us."

James Cox argues that, for Twain, laughter is the conversion of pain into pleasure, and Twain's own writings seem to support this hypothesis.[109] Twain argues in "Down the Rhone" that, just as an ice cube on your bare back cannot be told from a hot brand for a second or so, so can the shock or recognition that you have misperceived reality as a result of someone's performance of it feel like both pain and pleasure.[110] At first there is just the shock, and the question of how you respond to it is a question of social control. If you feel you were made, against your will, to misperceive reality by those who played on your assumptions to perpetrate an illusion on you, then anger and fear may result. However, your laughter can convert the shock of the experience to a reassertion of self-determination. It demonstrates a gap now between your self that was duped and your real, wiser self that can appreciate a good joke and will be harder to dupe the next time. Laughter creates distance between your old mode of perception and your new, enlightened self; anger and fear do not. That is what we mean when we say that people took something "personally." They were unable to create a distance between their negative face (how they see themselves) and their positive face (how they think others see them) through laughter.[111]

Twain knew all about this dynamic from his experience with frontier humor. Briefly, let us recap the social functions of the tall tale that bear on our discussion of Twain's social mechanics of laughter:

- The tall tale demonstrates control of unknown/frightening social and natural environment.
- The tall tale constructs a group of "outsiders" who fall for the tale and sets them apart from the "insider" or "insiders" who tell the tale.
- The tall tale criticizes the unwarranted (in the opinion of the insiders) elite status the outsiders enjoy and argues that their established traditions are useless on the frontier (cowboys vs. "city slickers").

Telling a tall tale was a powerful way to rebuild confidence, solidify insider status, and criticize outsiders. Being a victim of a tall tale left you with several options, the two primary ones being anger and laughter. Anger cemented your outsider status, and then you were left to other means—violence or the law—to challenge the insiders and retake control. Laughter, however, did not necessarily make you an insider, but it distanced you from the "outsider" values that the tall tale criticized. This laughter may not have felt at all pleasurable, despite Cox's argument, but it was a countermove to regain self-control against the social control exerted by the teller of the tall tale. That social control consisted of an awe-inspiring power to set community values and allocate community resources on the frontier. Insiders "got it," in more ways than one. Outsiders did not.

Twain made several important changes to this basic social machinery when he engineered his Petrified Man and Empire City Massacre hoaxes for the *Enterprise*. He was clearly familiar with Poe's hoaxes, as we saw above in the excised balloon chapter from *Life on the Mississippi*, and though he was born just a few months after the publication of Locke's Moon Hoax, he had undoubtedly heard of it by the time he worked on the *Enterprise*. Perhaps he even read it; the first reissue of the hoax, in an edition by William Griggs, was published in New York in 1852 just a few months before Twain moved to New York City. Twain knew that, in order to fly, hoaxes needed an "unfair pretense at truth" that the ephemeral orality of the tall tale could not quite produce. This "pretense" entailed appearing in print, in a news medium that people relied on exclusively for knowledge of what was going on outside their small western town. It also entailed the adoption of standard dialect and all the formal features encapsulated in readers' popsci. expectations,[112] as opposed to the vernacular dialect and narrative form of the tall tale. The print media offered both anonymity and distance that the telling of a tall tale could not; these mechanisms were crucial for slowing down any "facts" that might filter, through conversation between readers, into the workings of the hoax and grind its gears to a halt. A hoax also presupposed a level of publicity that the oral modality of the tall tale could not possibly achieve, thus increasing and speeding the diffusion of the hoax through multiple readerships and aggrandizing the reputation of its author.[113] In short, Twain knew a hoax worked its effect by pretending two things in one—pretending to be a "real" news story, thereby pretending that the events it reported had really happened. A tall tale simply pretended that the events it related were real, and sometimes it lost even that simple pretense with all its exaggerations for humorous effect.

A good tinkerer, Twain adapted the mechanics of the tall tale to suit his purposes. A tall tale, on the one hand, produced an awareness of outsider status and perhaps a chance to reassert self-determination through the distancing function of laughter. A hoax, on the other hand, could create something more. While the victim of a tall tale could learn little other than that she did not fit in on the frontier, the victim of a hoax could learn a lot more because his assumptions about science and the real world were the real target. This was Twain's design when he set out to "kill the petrifaction mania with a delicate, a very delicate satire" in the Petrified Man, and when he set out to expose the dividend-cooking mining companies in the Empire City Massacre. The hoax offers an opportunity to open readers' eyes to potentially dangerous assumptions they make about a particular social institution.

The hoax accomplishes this in the moment the reader perceives the gap, the lack, between what she has assumed and what the state of the art really is. It is a moment of embarrassment, in the "pregnant" root sense of the word—the moment when everyone else can see publicly what you know privately about yourself. If this moment of embarrassment is produced by an awareness of ignorance about a subject—science or economics, for instance, an educative potential to the hoax experience emerges. The distancing effect of laughter, in addition to reasserting self-control, can also express a desire for self-education as insulation against further attempts by others to control you through illusions. This instructive moment of embarrassment is very similar to the moment that Socrates's dialectic partners came to realize the gap of inconsistency between beliefs they entertained simultaneously—the *elenchus*.

What is particularly interesting is that there is some evidence that Twain was familiar with the *elenchus* as a dialectic method. While there is no direct evidence that he read Plato, he avidly enjoyed Voltaire, who channeled a great deal of Platonic dialectic into French commonsense criticism. Voltaire and Twain, in fact, are considered together as founders of the Freethinkers movement, an atheist movement beginning in the eighteenth-century that now is usually labeled "secular humanism." Voltaire's works contain dialectic passages similar to Socratic dialogues, where a naïve questioner is made by a master dialectician to admit the inconsistency of his assumptions. One of these passages can be found in *Candide* where Martin and Candide debate the jaded critical attitude of Pococurante, and Martin brings Candide to *elenchus* by pointing out an inconsistent belief he entertains, that Pococurante experiences pleasure by never having pleasure.[114] In *3,000 Years among the Microbes*, Twain constructs a very similar dialectic between Huck and a clergyman over the issue of animals having souls. The clergyman begins:

> "What is a creature?"
> "That which has been created."
> "That is broad; has it a restricted sense?"

"Yes. The dictionary adds, 'especially a living being.'"

"Is that what we commonly mean when we use the word?"

"Yes."

"Is it also what we always mean when we use it without a qualifying adjective?"

"Yes."[115]

And the dialectic continues until the clergyman forces Huck to see that he already believes all living beings, germs included, have souls because Huck does not distinguish "life" from "animation," or soul possession.

The question arises, then, if Twain was bringing his readers' understanding of the world to an *elenchus*, a null state of internal contradiction, why was he doing it? What did he hope to accomplish with his readers by entering into this dialectic with them in The Petrified Man? Joan Belcourt Ross argues that Twain wanted his readers to "see themselves true," to realize they were all the victims of hoaxes perpetrated not only by scientists, but by the church, by politicians, by Twain's favorite flogging horse, Christian Science, and so on.[116] Pascal Covici makes the same point in a slightly different manner: "The hoax-as-satire becomes especially important in Twain's works when it serves to reveal the hidden truth about the reader himself. Ourselves in particular, not people in general, are stripped of pretensions and made to stand self-revealed by Twain's most effective hoaxes."[117]

This goal as expressed by Ross and Covici is essentially a negative one. It argues for a deconstructive laughter that tears down the "wonder-business" of popular science but builds no belief structure in its place. This is the sort of laughter that follows a Poe hoax, as Poe had no vested interest in improving the lives of the people he fooled with his hoaxes.[118]

But Twain's hoaxes are not like Poe's. Twain states a social goal from the beginning of The Petrified Man, and that is to snap his readers out of their googly eyed fascination with paleontology while "touching up" the local coroner, to boot. If Twain was shaking up his readers' perceptions of reality, what did he want to put in place of their illusions, if anything?

Constance Rourke in her study of Twain's humor writes, "It is a mistake to look for the social critic—even manqué—in Mark Twain. In a sense the whole American comic tradition had been that of social criticism: but this had been instinctive and incomplete, and so it proved to be in Mark Twain."[119] Rourke, like Ross, sees Twain deconstructing the edifices in which his readers put their trust but refusing to inculcate them with his own values, tell them what to believe instead. In fact, however, Twain was explicit about what he thought Americans should believe. We have already seen his praise of working men over college boys in *The American Claimant*. He dedicated a book to exposing what he felt to be the fraudulent claims of Mary Baker Eddy in *Christian Science*. And he championed democracy against hereditary systems of power in *Huckleberry Finn*, *The American Claimant*, and *Connecticut Yankee*. In *Following the Equator*, he decried both missionaries and American Imperialism, arguing for the fundamental right to self-determination of all nations on earth.

This last was the social arena in which Twain was most active. He was vice president of the anti-Imperialist league from 1901 until his death in 1910 and a vocal opponent of both the Spanish American and the Philippine American wars, especially the brutal use to which technology was put in them. His essay for the *North American Review* in 1901, "To the Person Sitting in Darkness," was a scathing review of "the missionary question" in the wake of the Boxer Rebellion in China. In it Twain denounces the use of American technology to extend American dominance over less industrialized nations under the guise of missionary activity:

> Shall we? That is, shall we go on conferring our Civilization upon the peoples that sit in darkness, or shall we give those poor things a rest? Shall we bang right ahead in our old-time, loud, pious way, and commit the new century to the game; or shall we sober up and sit down and think it over first? Would it not be prudent to get our Civilization-tools together, and see how much stock is left on hand in the way of Glass Beads and Theology, and Maxim Guns and Hymn Books, and Trade-Gin and Torches of Progress and Enlightenment (patent adjustable ones, good to fire villages with, upon occasion), and balance the books, and arrive at the profit and loss, so that we may intelligently decide whether to continue the business or sell out the property and start a new Civilization Scheme on the proceeds?[120]

The essay created such a media firestorm, according to Jim Zwick, that one prominent Massachusetts editor claimed, "Mark Twain has suddenly become the most influential anti-imperialist and the most dreaded critic of the sacrosanct person in the White House that the country contains."[121] The backlash affected Twain significantly. What is today considered Twain's most powerful piece of antiwar literature, the "War Prayer," a black satire of the glories of war reminiscent of Stephen Crane's "War Is Kind," was considered so incendiary by Twain's biographer, Albert Bigelow Paine, that he urged Twain to suppress it; it was only published posthumously, during World War I.

Certainly, this late-life political cause against the connection between American technology and American Imperialism cannot be read back onto Twain's early hoaxing, like The Petrified Man, but it reflects the reaching of a state of critical mass of a concern that is present in that seminal hoax, present throughout all of Twain's writings—self-determination. If Twain had one absolute belief, consistently evident in his thinking, it was in the right of a human being to decide his/her own destiny. Twain's hoaxes gave readers a chance, through the social mechanics of laughter, to stand apart from their old preconceptions and choose a new path.

Lest we think of Twain as the champion of free thinking, however, it is important to point out that Twain also recognized the powerful mechanics of control inherent in the hoax. If the reader wished to be an "insider" like the author, then the hoax could instigate the kind of laughter that desired identification, an education in insiderhood, a distancing of the self from the old self-image, and a realignment with the values of the author and his insider group. Warwick Wadlington finds exactly this dynamic active in Twain's satiric travel narrative *Innocents Abroad.* In *The Confidence Game in American Literature*, Wadlington takes issue with James Cox, who says that Twain hoped to use his readers' laughter to "set the reader free" from her misconceptions. Wadlington finds instead that Twain wished to shatter his readers' habitual epistemologies and then reshape their worlds as Twain saw them. "The really pertinent test of sincerity in the book's rhetorical system is whether or not a given Twain performance accomplishes the twofold end of relieving excitable feelings and achieving authority over the reader."[122] Twain's hoaxes can be viewed as another rhetorical means to this end.

Twain in his last years had a clear view of the social mechanics of laughter. In "The Mysterious Stranger," he has Little Satan describe laughter as a technology—a powerful social weapon:

> [F]or your race in its poverty, has unquestionably one really effective weapon—laughter. Power, money, persuasion, supplication, persecution—these can lift at a colossal humbug—push it a little—weaken it a little, century by century; but only laughter can blow it to rags and atoms at a blast. Against the assault of laughter nothing can stand. You are always fussing and fighting with your other weapons. Do you ever use that one? No; you leave it lying rusting. As a race, do you ever use it at all? No, you lack sense and the courage.[123]

Twain may truly have wished for his readers to use this weapon to blast away the hoaxes being foisted on them by the American technological and imperial industries and "see themselves true," reassert control over their own lives. Or, he may have wished to use this laughter to break down his readers' value systems and replace them with his own—with the value of self-determination paramount over all. By times, maybe he used laughter toward both of those ends. But what is clear is that the social mechanics of the rhetoric he chose when he faced his readers in The Petrified Man were more than powerful enough to hit both of those targets.

Chapter Five

The Hoaxes of Dan De Quille: Building and Defending the West

Dan De Quille's scientific media hoaxing was deeply conditioned by both Poe's and Twain's hoaxes but was ultimately a different project from theirs. Poe did not have western pioneer readers—fiercely independent men and women who prided themselves on making up their own minds (sometimes independently of the facts), who were suspicious of outsider commercial interests, and who were in general ignorant of science but intimately familiar with mining technologies. And even though Twain and De Quille shared these readers, Twain could not relate to them at the level that De Quille did. De Quille admired his "stubborn old comstockers."[1] After all, he had been a miner himself for several years before turning to journalism, and he stayed in Nevada for almost thirty years after Twain went back East. His empathy with his readers permitted De Quille to simultaneously gull them and immortalize them as America's new folk heroes through his hoaxes. His pride in the character of the American prospector coupled with his love of science and technology made De Quille's hoaxing historically unique. Enamored by the tremendous potential of the hoax to construct realities, Dan De Quille used the four major and several minor scientific media hoaxes he wrote from 1865 to 1880 to create and defend his ideal West and westerners from eastern commercial exploitation.

Rhetorical Acculturation

Dan De Quille is somewhat of a mysterious character, only a fraction having been written about his life compared with the volumes penned on his famous colleague on the *Territorial Enterprise*, Mark Twain. Richard Dwyer and Richard Lingenfelter have produced the most recent and extensive biography of De Quille, amplifying Lawrence Berkove's excellent biography in his 1988 edition of De Quille's novella *Dives and Lazarus*.[2] In both of these portraits, De Quille appears as a cluster of contradictions: a devoted and supportive father and husband, who nonetheless left his family for nearly forty years to prospect and write in the West; a reputedly genial and nonconfrontational friend, who was also known to pick fights in bars for practically no reason; a dedicated journalist, by all contemporary reports the workhorse of the *Territorial Enterprise*, who was fired at least twice from that paper for being too drunk to work for weeks at a stretch; and a tolerant and well-read sociopolitical theorist, who turned out shockingly virulent anti-Semitic and anti-Chinese statements in his later life.

Dan De Quille was born William Wright on 19 May 1829, to a farming family in Knox County, Ohio. When he was eighteen, the family relocated to West Liberty, Iowa, and shortly thereafter, William's father died and left him largely responsible for his mother and eight younger siblings. At the age of twenty-four, he married Carolyn Coleman and had five children with her in four years, two of whom did not make it past infancy.

Little is known about Wright's early education other than evidence from his early letters that he was well lettered. He supposedly submitted stories to Eastern magazines, though there is no evidence that these were published.[3] From his letters we know that he read widely in world literature; some of his favorites included Don Quixote, many of Dickens's novels, the Arabian Nights, Ben Jonson, Jonathan Swift, James Fenimore Cooper, and Thomas Carlyle. He quoted frequently from Shakespeare and the Bible.[4]

In 1857 Wright left his family for the California/Nevada territories and gradually migrated to Virginia City in 1860, following rumors of new gold and silver claims. He would not return to Iowa, except for a few brief visits, for thirty-six years. He wrote his sister Lou religiously; if he wrote his wife and young children, those letters are no longer extant. Finally, his health broken, he returned to Iowa to live with his daughter in West Liberty for the last few years of his life.

De Quille never managed to strike it rich as a placer miner and supplemented his income writing for several territorial newspapers of the "Sagebrush School," including the *Golden Era* in San Francisco, the *Engineering and Mining Journal*, and, starting full-time in 1861, the *Territorial Enterprise*. It was about this time that he began to sign his articles "Dan De Quille." Unlike Twain he used this nom de plume exclusively until his death; most of his friends in the West knew him only by this name. His colleague C. C. Goodwin punned on De Quille's pseudonym in a sketch called "Dan and His Quills" that also provides us with a sampling of De Quille's journalistic repertoire:

> Across the table from us Dan De Quille is writing some of his abominable locals, and we have been studying his face. Of

> late he has thrown away his pencils and procured old-fashioned quill pens. . . . He writes ordinary locals with a turkey quill; for important affairs, like runaways and dog fights, he takes a goose quill; for obituary notices he keeps the plume of a raven; for mining reviews nothing will do but a swan's quill; his scientific articles are fashioned by the quill of an owl; while for the dreadful legends which he strings together for Sunday's *Enterprise* nothing will answer but a feather from the pinion of an eagle or an albatross.[5]

The Comstock, the region around Virginia City comprising the Comstock Lode and the towns that sprang up on it, provided a wealth of rhetorical opportunities for Dan De Quille. Contrary to many assumptions, Virginia City was not just a shantytown. Thanks to the largesse of citizens who had "struck it rich," the boomtown of nearly thirty thousand citizens possessed a full-fledged theater that put on Shakespeare plays, several musical venues, a newspaper vaunted as the best in the territory, a hotel that boasted the first elevator in the West, and social societies that hosted lectures and balls. Culturally and religiously, the Comstock was extremely diverse with Mormon, Catholic, and protestant Christian practices mixing and matching with the traditions of Jews, the Washoe and Piute Indians, and the large Chinese population who immigrated to the area to work on the railroads and in the mines.[6]

Newspaper clippings preserved in the Dan De Quille Papers at the Bancroft Library [BANC P–G 246], reveal that De Quille was curious about all of these different rhetorics and histories.[7] He clipped extensively on Asian culture, tucking into his scrapbooks a few postcards with pictures from Indonesia; he also collected articles on mythologies of several cultures, especially Native American cultures, and various religious and supernatural items. He also saved clippings on famous American authors, their personality traits, handwriting, and personal histories. A particularly telling clipping for our purposes is a memorial to Edgar Allan Poe clipped from the *New York Times* with mention of both Hans Phaall and the Balloon-Hoax.[8] These influences show up in both the topic and style of pieces De Quille wrote for the *Enterprise* and the *Golden Era*.

De Quille shared a room and a desk with Twain while working on the *Enterprise*, and they often helped each other out with stories—not just with topics, but also with argument structure and language. In general the object was humor and sensation, as the local columns were either filled with gunfights or the detailed history of a passing haywagon, and either extreme of interest had to be accommodated by an attention-arresting style in order to keep the subscription of the paper up. De Quille reminisced about trying to get Twain to leave more hints in Empire City Massacre that it was a sham and Twain waving him off with, "It is all plain enough."[9] It seems significant that De Quille, while he wrote many humorous articles, did not attempt a hoax until after Twain wrote the Petrified Man in 1862. And when he did, with the Silver Man for the *Golden Era* in 1865, it was on a very similar topic—a man found turned entirely to silver instead of stone. But where Twain left off scientific hoaxing after his one experiment, De Quille went on to produce four major hoaxes and at least twelve other short, humorous squibs, all concerned completely with science and technology. This predilection was in part due to De Quille's well-deserved reputation as one of the very first technical writers in the West.

Scientific Acculturation

De Quille claimed that Twain would have nothing to do with reporting science or geology at the *Enterprise* because he "hated to have to do with figures, measurements, and solid facts, such as were called for in matters pertaining to mines and machinery."[10] De Quille could not get enough of them. His papers are interspersed with back-of-the-envelope calculations of shaft depths and mine production rates and with sketched maps of the shafts of the Savage Mine and other big mines in the area.

Twain and De Quille did about equal time as placer miners. De Quille had no further formal scientific or technical education than did Twain. However, the difference between the writers that Judith Yaross Lee notes in her article "(Pseudo-) Scientific Humor"—namely that Twain makes fun of science like an amateur, and De Quille, like an expert—may have come down simply to a matter of interest.

De Quille was fascinated by all things scientific, and especially by mining. Most of his clippings are on geologic, metallurgic, and mining matters. He also kept up with news about climate and weather, chemistry, physics, astronomy, and "pseudoscientific" news about ESP and other psychic phenomena. His papers contain a bulletin from the Society for Psychical Research calling for news about hallucinations, thought transference, crystal vision, and automatic writing.[11]

In addition to keeping up with general scientific news, De Quille also wrote extensively and seriously on mining. Lawrence Berkove assesses De Quille's career as a technical writer as follows: "His reputation undoubtedly played an important part in establishing and maintaining the *Enterprise* as the dominant newspaper in mining circles. Even much later in his life, his articles on mining were solicited and published by a variety of periodicals, including specialized mining journals."[12] These journals included the *Mining Industry and Tradesman* and the *Engineering and Mining Journal*, for whom De Quille wrote not only histories and reports of the mines in the Comstock Lode, but also mining culture articles about "dowsing," or mineral divining, and the "Tricks of Miners." He was taken seriously by mining engineers at all levels. His clippings contain several favorable reviews of a proposal he apparently drafted for an improved method of constructing canals. His magnum opus, *The Big Bonanza: An Authentic Account of the Discovery, History, and Working of the World-Renowned Comstock Lode of Nevada*, was published in 1876 to universal accolades not only in the territories but also on the East Coast, and it is still considered the "bible" of Comstock mining history. De Quille's friend and colleague, editor Wells Drury, attested to De Quille's preoccupation with mining journalism:

> [H]is conscience never swerved from the firm conviction that the true calling of a first-class newspaper is to publish items concerning prospects, locations, mines and mills, shafts, tunnels, drifts, ore developments, stopes, assays and bullion outputs. All other matters to him appeared inconsequential and of no material interest. If there was a murder, a sensational society episode or a political contest, any of them were welcome to space after his mining notes were provided for.[13]

Aside from his technical mining reports, De Quille also wrote popular science news in at least three distinct registers: a "high style" formal technical manner, a "wonders of the world" style, and a humorous style. His high register can be typified by the following excerpt on assaying procedures from *The Big Bonanza*:

> In testing ores for silver, the miners in the early days used acids. . . . The heavy residuum was then washed from the horn into a matrass (a flask of annealed glass, with a narrow neck and a broad bottom). Nitric acid was then poured into the matrass until the matter to be tested was covered, when the flask was suspended over the flame of the candle or lamp and boiled until the fumes escaping (which are for a time red) came off white.[14]

While the language is as simple as possible aside from the mining jargon, and the argumentation is linear and logical, the passage lacks clever commentary or word play. It is this authoritative style that cemented De Quille's reputation as a trustworthy science writer, a reputation that became crucial to the reception of his hoaxes.

De Quille's middle register adopts a bit of "mystery" or "wonders of the world" rhetoric, as in this clipping from an article written for the San Francisco *Chronicle* on animal magnetism:

> Some remarkable discoveries have been recently made by French physicians in regard to what they call the action of medicines at a distance. . . .

> How were all these mysterious effects produced, often without even external contact? How could mercury blister the flesh through its tubes of glass and cloth envelopes? How could a medicine, placed unknown under a person's pillow, cause salivation, with the accompany symptoms? The substances were usually inclosed in paper, or in bottles, and many of them are odorless and cold send forth no effluvium to affect the patient's nerves. The whole matter is profoundly mysterious.[15]

De Quille's humorous register was easily identifiable, as in this job description for the state mineralogist of Nevada, published in the *Enterprise*: "He is to discover earthquakes and provide suitable means for the extermination of the same; also, for book agents, erysipelas, corn doctors, cerebro-spinal meningitis, and the Grecian bend."[16]

The impact of De Quille's many-layered scientific writing on his hoaxing is complex. Clearly, if he often wrote humorous pieces, his readers knew him to be just as capable of spinning a yarn as giving them a "true" report of what was going on in the mines or what new natural marvel had been discovered in the Nevada territory. To make matters worse, it appears De Quille regularly mixed modes in the same article. C. Grant Loomis writes in his extensive survey of De Quille's modes of journalism, "With no distinction between a true story or a fanciful one, he inserted the real and the false item into his daily public offering."[17] Twain and De Quille both practiced this "padding" in their frantic attempts to fill their local columns by press time. However, according to contemporary reports, this intermittent reinforcement only served to cement De Quille's reputation as a scientific "savant" with his readership. Attested De Quille's friend C. C. Goodwin, "what he wrote, everybody believed implicitly. This or that expert might make a report, and men would say, 'He may have been mistaken.' This or that owner of heavy shares might express his opinion, and men would say: 'Maybe his interests prejudice him.' But everyone believed Dan."[18]

In addition to his technical reputation, of course, De Quille's extensive experience reading and writing technical rhetoric stood him in good stead when he began his hoaxing with Silver Man in 1865. Judith Yaross Lee summarizes the effect of his rhetorical knowledge: "Dan De Quille knew enough science to fill his tales with incredible facts as well as convincing fantasies. In consequence, the stories conveyed an authentic respect for scientific knowledge in general and a persuasive pride in his own explanation of the 'truth.'"[19] As we will see, De Quille's hoaxes deliver a level of scientific "verisimilitude" in terms of both language and knowledge that rivals or surpasses Poe's. De Quille stuck to what he knew from experience with his hoaxes—mines, chemicals, and minerals. The prevalence of these topics in the everyday lives of his readers, coupled with his stand-up reputation as a science writer, made De Quille a formidable hoaxer.

De Quille's Hoaxes

Silver Man was De Quille's first "deliberate tall-tale creation."[20] G. Grant Loomis refers to these creations as "scientific tall tales" rather than hoaxes, an important difference in terminology that we will return to in the discussion section of this chapter. However, Loomis has made a careful assay of De Quille's scientific "sells" or "quaints," as the author referred to them. He found, in addition to the four major hoaxes considered in this chapter, twelve other minor squibs having to do with fantastic discoveries in geology, biology, or paleontology, all completely made up, most appearing in the *Territorial Enterprise* between 1867 and 1878. A complete list of these can be found in Loomis's article "The Tall Tales of Dan De Quille." In addition to Loomis's twelve, I have found a copy of one other hoax and mention of three more: the Mountain or Highland Alligator hoax, which drew a letter from the famous fossil collector Edward Drinker Cope; a hoax remembered by C. C. Goodwin having to do with the "excessive" water in the Comstock mines being an offshoot aquifer of Lake Tahoe; a hoax Wells Drury reported about a perpetual-motion windmill; and a hoax about a scientist hatching a live bird from a genetically engineered egg. In this chapter we will focus on De Quille's four most notorious hoaxes, along with Cope's letter about the Mountain Alligator.

Silver Man (1865)

As I have done in past chapters, I will use De Quille's first hoax to set up the major topics and issues concerned with his reading of reader expectations about science news. Then, we will consider his other major hoaxes and reader reactions to them before adjusting the filter of science reading expectations to reflect the milieu of De Quille's hoaxing. In some ways Silver Man is not the ideal hoax to begin with because, as C. Grant Loomis attests in his search for reprints in other papers after the publication date, "the story seems to have passed without any particular notice," which is not the case with De Quille's later hoaxes.[21] However, in this first hoax De Quille introduces all the rhetorical strategies he will continue to develop in his later hoaxes: an "Emperor's New Clothes" style of presuppository argument; emphasis on witness; argument for plausibility via analogy; and skillful exploitation of the codependence of doubt and belief. He also uses a unique form of *refutatio* in which he monitors his readers' interpretive process on-line, so to speak, and adjusts his arguments to cater to what he believes to be their highest-ranked expectations.

"The Wonder of the Age: A Silver Man" appeared in two long columns on pages three and four of the 5 February 1865 edition of the *Golden Era*, the San Francisco paper that first employed De Quille after his arrival in the territories. Two advertisements for the piece appeared on page one: "The Marvelous 'Silver Man' is described by Dan De Quille in another column"; and "Dan De Quille, the Sage Brush Humorist of Silver Land, discourses on a scientific subject with the spirit of a true savan." While these announcements have a coy tone, they may, first of all, have gone unnoticed by readers, sandwiched as they were down in the lowest columns of the first page; second, they may not have prejudiced readers against the truth of De Quille's argument, since as argued earlier, he had been a contributor of mining news to the *Golden Era* for five years at this point, and his readers knew that he wrote both serious and humorous scientific news. Therefore, readers were likely committed to judge for themselves at this stage, as De Quille himself expected them to do.[22]

The entire story is too long to quote here, but it concerns a man found turned entirely to silver—all the way down to silver pyrite crystals encrusting the cavities in his bones and between his garments—down in the "Hot Springs Lead" deposit in a local mine. De Quille begins with an argument for the credibility of the finding in spite of its incredible appearance. He goes on to cite the individuals involved in the discovery, to give the history of the lead deposit and the mine in which it was found, and to relate the details of the discovery itself. The article finishes first with a statement that many witnesses have seen the "Silver Man" (which, unfortunately, was deteriorating so rapidly from oxidation that it would not be viewable much longer) and with a lengthy analogy to two similar findings, one a Swedish "copper man" in a mine, and one the finding of a French chemist.

Immediately, we recognize many familiar features of the hoax: the "hot topic" of a mineralized human being in the midst of daily finds of fossils and petrifactions; the plausibility of the well-known location and the "Hot Springs Lead" deposit; the ethos leant by the witnesses and the foreign scientists; and the minute details of the mineralization process. However, De Quille uses some rhetorical strategies we have not analyzed in previous hoaxes. We will consider each in roughly the order they are employed in the hoax and ask how the choice of this strategy reflects De Quille's mental model of his readers' expectations.

The "Emperor's New Clothes" Ploy

De Quille opens the hoax with a presupposition that the silver man exists and his readers are already familiar with it: "Dear Era:—Everybody, no doubt, has heard of the discovery of the wonderful "Silver Man," found in a mine between Esmeralda and Owen's River."[23] The use of the definite article to introduce "the discovery" and "the silver man" here, in contrast to the indefinite "A Silver Man" of the title, adds linguistic force to a presupposition of existence and publicity for the discovery. Later in the story, De Quille employs a similar strategy of presupposition. "All who have the least knowledge of palaeontology know that all those wonderful remains of fishes, animals, and so on, found in limestone and other rocks, and about which so much is said and written, are not the creatures themselves, but merely their shapes replaced by mineral substances."[24] De Quille, here, is constructing an audience of savants by using an "anyone who disagrees with these facts is necessarily ignorant" line of argument.

The pressure on De Quille's readers of this strategy is not immediately apparent unless we return to the self-sufficient Jacksonian pioneer that played the foil to the dupe of many a tall tale in Twain and De Quille's era. Just as the tall tale depended for its success on hapless outsiders trying to make a show of familiarity with their bizarre new environment in the West, De Quille is trying to force his pioneer readers to acquiesce to his argument by making the alternative to belief unpleasant—looking to everyone else like an unsavvy outsider. It is the "Emperor's New Clothes" strategy, where even though the emperor's senses tell him he is naked, he would rather appear naked than pas au courant. Similarly, De Quille's pioneers would rather risk jumping to the wrong conclusion than being labeled an outsider.

Figure 5.1 Dan De Quille (William Wright). *Courtesy, Special Collections, University of Nevada–Reno Library*

Support for De Quille's attraction to this strategy comes from Neil Harris's analysis of contemporary responses to P. T. Barnum's hoaxes. Harris explains, "Men priding themselves on their rationalist, scientific bent, familiar with the operation of novel machines, aware of the variety of nature, tended to accept as true anything which seemed to work—or seemed likely to work."[25] De Quille's pioneers grew up in the same exciting era of industrial genesis and possibility as Barnum's viewers; most of them had lived in that technologically charged eastern environment just a few years prior to reading De Quille's articles. Accordingly, they may have felt the enormous social pressure that came with the Jacksonian territory—the pressure to appear self-sufficient or risk seeming un-American. Therefore, in situations where they knew themselves to be undereducated, De Quille's readers may have thought it best to make a show of competency. To corroborate Harris's theory, I examined over two dozen interviews with gold rush pioneers from De Quille's time that were recorded by Hubert Howe Bancroft's History Company. In many of these interviews, self-reliance and the overcoming of obstacles "on one's own hook'" were two persistent themes as pioneers reported their life philosophies.[26] De Quille seemed to be counting on just this strong drive to appear independent and competent when he employed the "Emperor's New Clothes" rhetoric to coerce belief in his hoax.

Figure 5.2 Nevada Miners. *Courtesy, Special Collections, University of Nevada–Reno Library*

Refutatio

With the "Emperor's New Clothes" strategy, De Quille shows himself sensitive to what he believes is his readers' desire to appear on top of the latest developments in science (related to their novelty expectations). With his unique *refutatio* strategy, he makes an argument that plausibility is even more important to his readers. Almost immediately in Silver Man he injects a rebuttal argument (again, I have bolded terms associated with expectations and underlined conflict markers):

> Everybody, <u>however</u>, has not heard the full particulars of the discovery, and many will <u>hoot</u> the idea of any such **discovery** ever having been made.
>
> They will at once say that it is **impossible** for a human body to be changed to silver ore—Let them have their say!
>
> <u>Although</u> the story is **almost too much for belief**, <u>yet</u> I hope to be able to show, before finishing this account, that, **startling** as the assertion may appear, such a change in the substance of the human body is not only **possible**, but that there is on record one well **authenticated** instance of a similar changing of a human body into a mass of ore.[27]

Here, De Quille constructs readers who are not impressed by the sensation and novelty of the silver man but instead "hoot" at its "impossibility." He models for himself a "false" interpretive judgment they might have made in the face of the conflicting expectations of novelty/sensation and plausibility. To try to correct this "false" impression, De Quille reassures his readers he will indeed offer arguments through logic and real-world analogy that will satisfy their high-ranked expectation of plausibility (thus removing the * under that expectation) and will provide corroboration (thus adding a violation to the "false" interpretation; that is, if they wanted to disbelieve the story at that point, readers would have to overlook the corroboration De Quille would provide). Thus, his *refutatio* is a metarhetorical strategy, an attempt to persuade by intervening in the interpretive process—just as his hoax is an attempt to persuade through intervention in the process of science

Table 5.1 Reader Reaction to Silver Man Projected by De Quille's *Refutatio*

	Plausibility	Corroboration	Novelty	Sensation
TRUE	−(*)			
✓FALSE		+(*)	*	*

Figure 5.3 Composing Room at Territorial Enterprise showing desk used by De Quille and Twain. *Courtesy, Special Collections, University of Nevada–Reno Library*

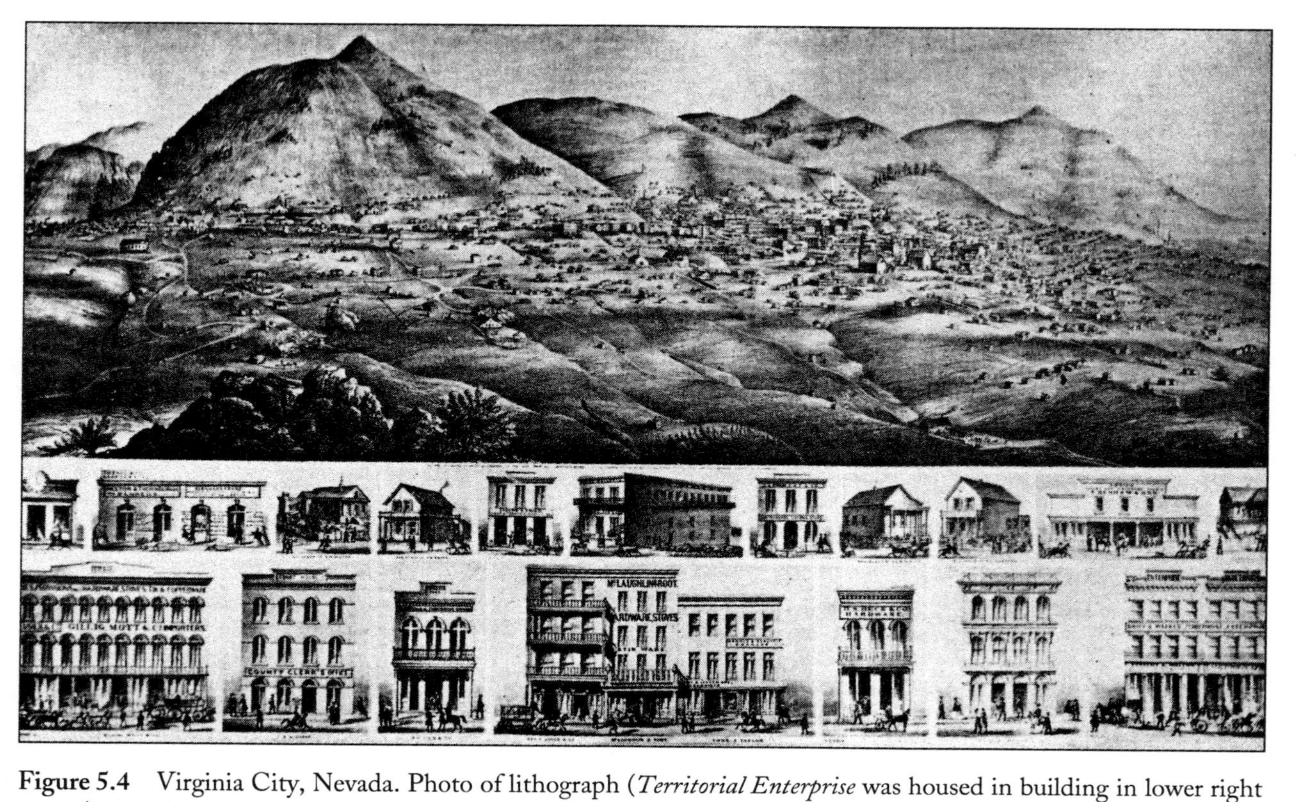

Figure 5.4 Virginia City, Nevada. Photo of lithograph (*Territorial Enterprise* was housed in building in lower right corner). *Courtesy of Special Collections, University of Nevada–Reno Library*

popularziation. De Quille immediately proceeds to make an interesting variation of the "mystery" move from our antebellum hoaxes:

> We have had all kinds of astonishing discoveries. Many things formerly classed among the impossibilities are now familiar, every-day possibilities. We are now to acknowledge that it is not impossible for a human body to be changed—through contact in a mineral vein with solutions of certain salts, carbonic and hydrosulphuric gases, and the electrical currents induced by the reaction of said solutions upon each other—into a mass of sulphuret of silver. (pg. 3)

Instead of the traditional "mystery" opening, this is an "antimystery" argument about the power of science to demystify what has previously seemed awesome and supernatural. This rhetorical strategy is a marked change from the rhetoric of the 1830s where science and mystery cooperated. De Quille's appeal to scientific fact and jargon (plausibility, popsci., and detail) are time-honored strategies. However, this time they are meant to reassure, not to overwhelm, his readers as miners were familiar with these compounds from the reactivity tests they routinely performed on ore to determine if it contained silver.

It appears from this *refutatio* strategy that De Quille believed himself to be dealing with a readership that was less swayed by novelty than by plausibility. This conclusion is borne out by his next strategy, which is the establishment of multiple witnesses to the discovery.

Eyewitness as Corroboration

De Quille provides copious witnesses to the silver man, but these do not exploit the same sort of ethos provided by the 1830s references to corroborating journals or by the name-dropping of foreign scientists. Instead, Kuhlman and the miners who discover the silver man in De Quille's story, along with the "scientific friend" of Kuhlman in Aurora, the viewers of the silver man, and De Quille himself, who saw a button of silver produced from the silver man—these locals provide a vicarious experience of witness for the hoax's readers.

Newsreaders counted on journalists to bear witness to things the readers themselves did not have access to, and thus to construct for them much of the world that spread beyond the realm of their senses. This was doubly true in De Quille's milieu, where pioneers did not have readily available to them independent sources for confirming or denying the truth of what was printed in the newspapers. This coerced trust between reader and journalist helps explain some of the personal animosity that is still directed toward hoaxers today when their hoaxes are exposed. What is at stake is not just the journalist lying; the world the reader has been living in suddenly does not hold true. De Quille seemed to sense the importance of vicarious witness because he anchored the truth of the "silver man" to the experiences of character types his readers would trust—local miners, journalists, and scientists. He respects the high ranking of corroboration left as a legacy from antebellum newsreading, but for his western readers, it morphs into witness.

Analogy

Throughout the Silver Man, but especially in the later sections, De Quille makes arguments for the plausibility of the silver man by analogy. He harps at some length upon the analogy of several Swedish miners found turned to copper. (Odds are good he fabricated this analog, making the Silver Man in actuality a sort of "marushka doll" of hoaxes.) He also mentions, as quoted above, the fossilization of fishes and other animals, a process similar to the mineralization of the silver man, which De Quille knew his readers would find familiar. He finishes with an analogy to crystallization processes studied by M. de Senarmont, an actual French chemist.

Argument by analogy was one of the original components of popsci. However, analogy takes on additional significance given a new reading context: the conditions of life on the frontier. Anyone who has spent any time in the West knows (!) that the majority of place names are analogies to forms or places in the homelands of the pioneers; thus, you have New England Glasgow, Montana, and Paris, Texas. Geologic forms are also usually named for what they resemble—camels, wagons, or breasts.[28] Naming by analogy is a way to reduce the complexity and threat of a foreign environment. By relying on analogs as a central proof in the Silver Man, De Quille counted on his readers to use the same epistemological strategy when considering his hoax as they did in their everyday lives on the frontier—and to use it to his advantage.

Codependence of Belief and Doubt

De Quille makes a move in Silver Man that seems odd at first if we assume that his first priority is to present a seamless hoax. He acknowledges his readers' doubt by admitting his own, albeit on minor points of the narrative. In describing the silver man, De Quille writes, "The body is supposed to be, and doubtless is, that of an Indian; but in its present changed state it is impossible to be certain on that point."[29] He reintroduces the impossibility of proving the "Silver Man" a bonafide artifact at the end of the story: "I might say much more in proof not only of the fact of a human body so changed having been found, but of the simple and natural causes which have operated to produce a change which at the first glance appears so wonderful; however, as many would not believe, even though I should produce the body and melt it up into buttons before their very eyes, I refrain."[30]

De Quille's rhetoric of doubt and belief here conflates the two at every turn. First, he uses the word "doubtless," which introduces his readers' "doubt" even as it implies they should be sure of what De Quille reports. He argues twice for the impossibility of proving the silver man genuine, but the somewhat sarcastic reason he provides is that some of his readers would doubt no matter what ironclad proof he offered them. In this way, De Quille cleverly constructs a believing readership on the "Emperor's New Clothes" paradigm again, this time by providing a negative model for their interpretive behavior: in paraphrase, "Only ignorant readers masquerading as skeptics could still doubt this story." No one wants to be in that crowd. Because there is no extant reader reaction to the Silver Man hoax, we cannot gauge the extent to which readers responded to De Quille's manipulations of belief and doubt in this hoax, but De Quille's focus on these techniques testifies to his belief, as a member of his own readership, in their effectiveness.

The motions of concealment and exposure, and the motions of belief and doubt, share the same epistemological arc between the unknown and the judging mind, between the data of the senses and the evaluation of that data. On the receiving end, that arc could be described just as De Quille defined the responsibility of his readers, "deciding for themselves." On the authorial end, that arc is the sheer pleasure of using words to construct realities for a public kept constantly at a distance by those words from first-hand experience of those realities. This was De Quille's game, and he refined it over the next thirty-five years through the publication of three more major hoaxes.

Solar Armor (1874)

Entitled "Sad Fate of An Inventor," a story appeared in the 2 July 1874 edition of the *Enterprise* describing an invention gone horribly awry:

> A gentleman who has just arrived from the borax fields of the desert regions surrounding the town of Columbus, in the eastern part of this State, gives us the following account of the sad fate of Mr. Jonathan Newhouse, a man of considerable inventive genius. Mr. Newhouse had constructed what he called a "solar armor," an apparatus intended to protect the wearer from the fierce heat of the sun in crossing deserts and burning alkali plains. The armor consisted of a long, close-fitting jacket made of common sponge and a cap or hood of the same material;

> both jacket and hood being about an inch in thickness. Before starting across a desert this armor was to be saturated with water. Under the right arm was suspended an India rubber sack filled with water and having a small gutta percha tube leading to the top of the hood. In order to keep the armor moist, all that was necessary to be done by the traveler, as he progressed over the burning sands, was to press the sack occasionally, when a small quantity of water would be forced up and thoroughly saturate the hood and the jacket below it. Thus, by the evaporation of the moisture in the armor, it was calculated might be produced almost any degree of cold. Mr. Newhouse went down to Death Valley, determined to try the experiment of crossing that terrible place in his armor. He started out in to the valley one morning from the camp nearest its borders, telling the men at the camp, as they laced his armor on his back, that he would return in two days. The next day an Indian who could speak but a few words of English came to the camp in a great state of excitement. He made the men understand that he wanted them to follow him. At the distance of about twenty miles out into the desert the Indian pointed to a human figure seated against a rock. Approaching they found it to be Newhouse still in his armor. He was dead and frozen stiff. His beard was covered with frost and—though the noonday sun poured down its fiercest rays—in icicle over a foot in length hung from his nose. There he had perished miserably, because his armor had worked but too well, and because it was laced up behind where he could not reach the fastenings.[31]

This hoax takes the form of a full-fledged modern news article. It starts with a summary of the "sad fate" of the inventor and then moves to the details of the story—the who, what, when, where, why, and how. It finishes with an analysis of how the death must have transpired. Terse and to the point, it avoids a great deal of the rhetorical flights of the "mystery" openings and "benefits to mankind" conclusions that the 1835 hoaxes and science articles exhibited. As evidenced by De Quille's science writings about dowsing and other "mysterious" phenomena, science now provided natural explanations for many things that had theretofore seemed supernatural or mysterious. This may account for the stripping of the grandiose language from hoaxes written after 1865. However, science and nature also presented pioneers with many experiences and objects that were beyond the pale of their experiences back East. De Quille definitely exploited this "supernatural" aspect of scientific inquiry when presenting a man freezing to death in a roasting desert.

De Quille includes many scientific details about the solar armor that would satisfy readers' detail expectations as well as their plausibility expectations, such as the careful description of the pump apparatus. As Poe did with his balloon, De Quille walks readers through the workings of the pump rhetorically to actually create a working pump in their imagination. He adds to these appeals the appeal of local eyewitnesses in the forms of prospectors, again, and an Indian. Readers' novelty and sensation requirements are certainly satisfied in the irony of death by freezing in Death Valley.

The most interesting aspect of this hoax, however, is its reprinting history. The story was copied widely, and we have many of these reprints thanks to De Quille scrapbooking them and mentioning specific reprinting papers and dates. He wrote to his sister Lou Wright on 23 August 1874, "My story of the man who was frozen to death by a solar armor of his own invention was illustrated in one of the Eastern pictorials. It was not well done, however. The artist made a horrible looking beast of poor Woodhouse. The Scientific American thought enough of that sell to copy it, it being somewhat in their line."[32] The *Scientific American* reprint De Quille refers to here appeared on 25 July 1874. The editors' introduction to the story, titled "Sad Fate of a Nevada Inventor" was rather coy: "The coolest and most refreshing item we have read since the commencement of the heated term lately appeared in the Virginia City (Nevada) *Enterprise.*"[33] However, the editors went on to reprint the story almost verbatim, and it appeared not on the first page, where jokes and anecdotes usually appeared, but on page 51 alongside engravings of a new "Apparatus for Transplanting Trees" as well as a description of "A New Alkaloid from Morphine" and an innovative air-conditioning system used in the House of Commons in London.

We know the story was also reprinted in several New York papers, including the *Sun*, for which De Quille was a regular western correspondent. De Quille's colleague at the *Enterprise*, C. C. Goodwin, reported that De Quille received a copy of the London *Times* in the mail with a copy of the Solar Armor story in it, including recommendations by the editor that the British army consider the armor for equipping its soldiers in India and other hot climates.[34] Goodwin claimed that De Quille bracketed the story with an elaborate picture of a man thumbing his nose (reminiscent of the posture of Twain's Petrified Man!) and mailed the paper back to the English science writer. Delancey Ferguson discredits Goodwin's report because, according to his research, the hoax never made it to the *Times.*[35] However, Ferguson checked the *Times* for 1862, the year of Twain's hoax, and not 1874, the year of the Solar Armor hoax. The story does indeed appear in the London *Times* on 27 July 1874 under the title "Too Successful" with no editorial criticism. However, this article makes no mention of using the solar armor for British troops.

The London *Daily Telegraph* for 3 August 1874 reprinted the story, and it is worthwhile considering the editor's comments because they become the catalyst for De Quille's next installment of the hoax. The *Daily Telegraph* editor prefaces the story with a brief description of Virginia City and the scalding desert to the east of it, which makes "men—and even wagons, with their teams of from eight to sixteen mules or oxen—to sink overwhelmed with heat and thirst when an effort is made to cross this desert in summer." This description seems to validate Newhouse's ultimately tragic quest for a "solar armor" to combat the heat. However, the *Daily Telegraph* finishes with this evaluation (expectations are bolded and conflict markers underlined):

> The **marvelous** stories which **come from "the plains"** are apt to be received with incredulity by our transatlantic kinsmen who dwell upon the Eastern seaboard of the United States. We confess that, although the fate of Mr. Newhouse is related by the Western journal ***au grand serieux***, we should require some **additional confirmation** before we unhesitatingly accept it. But every one who has iced a bottle of wine by wrapping a wet cloth round it and putting it in a draught, must have noticed how great is the cold that evaporation of moisture produces. **For these reasons** we are disposed to accept **the tale from Virginia City** in the same frame of mind which Herodotus, the Father of History, usually assumed when he repeated some **marvel** that had reached him—that is to say, we are neither prepared to disbelieve it wholly nor to credit it without question.[36]

The editor for the *Daily Telegraph* lets us glimpse the competition between his expectations that leads to a suspension of judgment. First, he cites the "marvelous" quality of the story in a negative light. His comments may reference a shift in attitude about the sensation expectation since 1835: namely, as the English have gained far more control over their environment through science and technology than have American pioneers, sensation might actually acquire a negative valence for the English popular scientific press, since sensational events resist the control and logical consistency of a scientific epistemology. In addition to the negative light of "marvelous," the editor cites the dubious authority of newspapers from "the plains" as a weak point of the story (medium). However, he goes on to counterbalance these negative points with the conformity of the story to popsci. expectations, "*au grand serieux*." He adds to the positive column a supporting analogy of cooling by evaporation (plausibility). "For these reasons," the writer concludes, the *Daily Telegraph* suspends its judgment until it receives "confirmation" (corroboration).

Table 5.2 Editor's Suspended Judgment about Solar Armor Hoax

	Corroboration	Sensation (−)	Plausibility	Medium	Popsci.
TRUE		*		*	
FALSE			*		*

This is the point at which the common reading filter we have developed really starts to work for us, making clear why the editor cannot resolve this complex decision. The editor's interpretive activity can be represented graphically as in table 5.2. Assuming the editor ranks his expectations roughly as the common reading filter updated at the end of the previous chapter would predict, we can clearly see his dilemma. Medium and popsci. conflict with each other, and they are the same level of rank; the same dynamic holds for sensation and plausibility (I am experimenting with negative-polarity sensation [−] here: "Sensational science news stories are usually not true."). That is why the editor indicates that he needs corroboration to make his decision; it is a higher-ranked constraint and will decide the "locked" contest.

De Quille used this editor's doubts as a springboard to launch the second installment of his hoax, "A Mystery Explained.—The Sequel to the Strange Death of Jonathan Newhouse, the Inventor of the Solar Armor." The sequel, a strategy perhaps suggested by Mark Twain's sequel to The Petrified Man, appeared in the *Enterprise* on 30 August 1874. In it De Quille casts the *Daily Telegraph* editor in the role of a doubter:

> [A]s the truth of our narration appears to be called in question, if not directly at least impliedly, by a paper which enjoys the largest circulation of any daily newspaper in the world, we feel that it is but right that we should make public some further particulars in regard to the strange affair—particulars which throw a flood of light upon what, we must admit, did appear almost incredible in our account of the sad occurrence as published. It seemed strange that so great a degree of cold could be produced simply by the evaporation of water, but it now appears that it was not water—at least not water alone—that was used by the unfortunate gentleman.[37]

De Quille aims to give the editor exactly what he wanted: corroboration and plausibility. He first summons a witness, David Baxter, the coroner at Salt Wells, who performed an inquest on the body of Jonathan Newhouse; he also lists all of Newhouse's statistics, including place of birth.

The text of the inquest, inferred rather than quoted, is a metonym De Quille uses to corroborate the Solar Armor; i.e., the coroner's report exists and is true, so by extension the Solar Armor must exist. De Quille would develop this strategy of textual witness even further in his last two hoaxes.

Next in the sequel, De Quille produces more witnesses—prospectors at the camp from which Newhouse began his fateful walk. To boost the tale's plausibility, De Quille reports that these men found a satchel belonging to Newhouse that contained chemicals that intensified the evaporation within his solar armor. Several of the witnesses reported frostbite on their hands from handling the body. The coroner supposedly tells De Quille that he is sending the chemicals on to the Academy of Sciences at San Francisco (a genuine organization) for analysis. De Quille wraps up, however, with his signature performance of doubt. "Whether or not he has done so we cannot say. For several weeks we have closely watched the reports of the proceedings of the learned body named, but as yet have seen no mention made of either the chemicals or the armor."

This article spawned a new flurry of reprinting on the East Coast as papers responded both to the *Daily Telegraph* criticism and to De Quille's new installment. The New York *Sun* defended itself for having published the first installment in good faith by now reprinting the "additional proof" of the coroner's report on 17 September 1874.[38] The New York *World* took a slightly cagier approach. It started off, "About two months ago the Virginia City *Enterprise*, of Nevada, a journal that so ingeniously mingles fact with fiction that its readers are never weary of exclaiming, 'Well, I wonder!' related the incidents connected with the demise of Mr. Jonathan Newhouse." The *World* next juxtaposed the *Telegraph*'s complaints with De Quille's rebuttals, ending with an tongue-twisting list of the chemicals from the "inquest": "'Ammonic nitrate,' 'sodic nitrate,' 'Ammonic chloride,' 'sodic sulphate,' and 'sodic phosphate.'" The article at last concluded, "Let the *Telegraph* now fold its hands, murmur '*si non e vero*,' &c., and be satisfied.'"[39] *Si non e vero, e ben trovato* translates, "If it's not true, it's well done/worked" and is reminiscent of Poe's claims that if his inventions did not actually exist and work, it was not because they *could* not. The *World* thus seems to take De Quille's sequel as demonstrating a level of proof beyond which it is fruitless to argue, as what would be required after that would not be more violation-producing data, but actual reranking of expectations, an action unlikely to occur on the basis of reading a single story.

Overall, the second wave of debate about the Solar Armor hoax focused positively on the testimony of the coroner and the further scientific justification De Quille provided. This reaction confirmed the efficacy of the strategies of corroboration and scientific plausibility that De Quille was refining through this hoax. Also, the editors of the reprinting papers focused on the fact of De Quille's sequel itself as authoritative proof of the verity of the Solar Armor, providing powerful reinforcement for De Quille's desire to franchise his other hoaxes, most notably the Traveling Stones hoax, which boasted a public half-life of twenty-five years.

Traveling Stones (1867, 1876, 1879, 1892)

The Traveling Stones hoax was De Quille's most famous and longest running. The story was about magnetic stones that would move from wherever they were scattered to cluster together. It first appeared in the *Enterprise* on 26 October 1867 and was reprinted in his history of Nevada mining, *The Big Bonanza*, in 1876. Here is the version that originally appeared in the *Enterprise*:

> A gentleman from the southern part of Pahranagat, who passed through this city a day or two since on his way to Sacramento, Cal., showed us a half a dozen or so of very curious pebbles—not curious in appearance, but rather curious in action. They were almost perfectly round, the majority of them nearly as large as a black walnut, and appeared to be of an irony nature. About the only remarkable thing about these pebbles—and it struck us as rather remarkable—was that when distributed about upon a floor, table or other level surface, within two or three feet of each other, they immediately began traveling toward a common centre and there huddled up in a bunch like a lot of eggs in a nest. A single stone, removed to the distance of three and a half feet, upon being released at once started off with wonderful and somewhat comical celerity to rejoin its fellows; taken away four or five feet it remained motionless. Mr. Hart, the gentleman in whose possession we saw these rolling stones, says they are found in a region of country that, although comparatively level, is nothing but bare rock. Scattered over this barren region are little basins, from a few feet to a rod or two in diameter, and it is in the bottom of these that the rolling stones are found. They are from the size of a pea to five and six inches in diameter. The cause of these stones rolling together is doubtless to be found in the material of which they are composed, which appears to be loadstone or magnetic iron ore.[40]

In spite of the wink-and-nudge use of the word "irony" in the second sentence, De Quille expends some effort in making the stones sound genuine. The best lie, as they say, is one that has a lot of the truth mixed in, and De Quille got his inspiration for this hoax from actual geologic discoveries. Among the newspaper clippings in his papers at the Bancroft is an article on the "seven wonders of Corea including a floating stone" and a "hot stone glowing on top of a high hill."[41] He also clipped an article entitled "Are Stones Alive?"[42] The hoax also contains many familiar features including a trustworthy local "gentleman," the mystery language of "very curious" (although the opening of the piece is still straightforward "who, what, when, where, why" rhetoric), and a scientific explanation of the stones' behavior due to their composition of "loadstone or magnetic iron ore."

This hoax apparently worked spectacularly well. Contemporary Wells Drury describes De Quille getting an offer of ten thousand dollars for a "few bushels" of the stones from P. T. Barnum, on the condition that they could be made to perform under the big top. De Quille also supposedly received a request from a German physics society to examine the stones.[43] These requests are not extant in De Quille's papers, but neither allegation is unlikely, as we know that De Quille's previous stories were reprinted in European papers and that Barnum was always on the hunt for new attractions to add to his shows.

The rhetoric of this initial phase of the hoax is relatively pedestrian, but its rhetorical history is fascinating, for De Quille, encouraged by the success of his sequel to the Solar Armor hoax, added two more installments to the Traveling Stones hoax over the next twenty-five years. The next installment appeared in the *Enterprise* on 11 November 1879 (quoted from Dwyer and Lingenfelter):

> [T]he story of the little traveling stones seemed to supply a want that had long been felt—to fit exactly and fill a certain vacant nook in the minds of men—and they traveled through all the newspapers of the world. This we did not so much mind, nor were we much worried by letters of inquiry at first, but it has now been some years since we ceased to enjoy them. First and last, we must have had bushels of letters asking about these stones. Letter after letter have we opened from foreign parts in the expectation of hearing something to our advantage—that half a million had been left us somewhere or that somebody was anxious to pay us four bits a column for sketches about the mountains and the mines—and have only found some other man wanting to know all about those traveling stones.
>
> So it has gone on all these fifteen years. Our last is from Tiffin, Ohio, dated Nov. 3, and received yesterday. His name is Haines, and he wants to know all about those stones, could he obtain several and how? Not long since we had a letter from a man in one of the New England States who informed us that there was big money in the traveling stones. We were to send him a carload, when he would exhibit and sell them, dividing

> the spoils with us. We have stood this thing about fifteen years, and it is becoming a little monotonous. We are now growing old, and we want peace. We desire to throw up the sponge and acknowledge the corn; therefore we solemnly affirm that we never saw or heard of any such diabolical cobbles as the traveling stones of Pahranagat— though we still think there ought to be something of the kind somewhere in the world. If this candid confession shall carry a pang to the heart of any true believer we shall be glad of it, as the true believers have panged it to us, right and left, quite long enough.[44]

De Quille seems to derive no little pleasure from the "pang" his revelation is causing the eastern businessmen and other believers in his hoax, although no evidence of these "bushels" of letters remains other than Drury's testimony to Barnum, the "man in one of the New England States," offering to pay ten thousand dollars for the stones. Interestingly, however, De Quille backs away from disavowing the stones' existence altogether, hinting that he might have been prescient rather than a flat-out liar. This hint will turn into a strategy of hoax perpetuation in the final installment of the Traveling Stones hoax. Notice also another innovation De Quille is making with the corroboration expectation. Instead of merely mentioning eyewitnesses, as he did in the Silver Man hoax, De Quille is now citing textual evidence, "letters" from other believers. The mere mention of these texts serves to reify the existence of the traveling stones in readers' imaginations.

On 6 March 1892, a story called "The Traveling Stones of Pahranagat" appeared in the Salt Lake *Daily Tribune* as part of an article entitled "Undesirable Thriftiness." In the story, De Quille explained that he had written his original Traveling Stones hoax with another object, in addition to teasing his readers, in mind; he had put it out as a "feeler." He claimed, "My object was to set the many prospectors then ranging the country to looking for such things." He went on to state he had confessed the hoax only out of exasperation with all the requests he received for the stones. From there, the article took an interesting turn:

> Shortly after I denied the existence of the traveling stones, I began to receive assurances that such stones had really been found in central Nevada. Among others who had found and owned such stones was Joseph E. Eckley, present State Printer of Nevada. Mr. Eckley has several times told me of his having owned a lot of such stones which he was a citizen of Austin, Lander county. He obtained them in Nye county on a hill that was filled and covered with geodes. Most of these geodes contain crystals of various colors. These are not the traveling kind.

> Those that appear to be endued with life are little nodules of iron. They are found on the hill among the geodes, and it was only by accident that Mr. Eckley discovered their traveling propensities. He had the stones he found for some months, and frequently exhibited them. This finally led to their being stolen, some one breaking open his cabinet and carrying them off. Mr. Eckley is a truthful man. He now resides in Carson City, and doubtless would be able to give further particulars in regard to the stones he discovered.

Not satisfied with this single appeal to eyewitness, De Quille went on to reprint a "letter" from another prospector in Idaho who claimed to have found traveling stones in Humboldt County and offered to go get them so Nevada could present them at the next World's Fair. His citation of textual authorities to satisfy readers' desires for vicarious witness has at this stage turned into full-blown forgery of these supporting texts. He is expanding on the "marushka doll" strategy of embedding hoaxes within hoaxes that he began by inserting the fake analogy of the Swedish "copper man" in the Silver Man hoax.

A more pressing question about this last installment of the Traveling Stones hoax is the following: why would De Quille revive this hoax after having exposed it thirteen years before? His motivation is impossible to reconstruct, but his arguments in the last installment of the hoax lead to some suggestive conclusions that show him developing other strategies nascent in the Silver Man hoax. In addition to the striking development of corroboration, this last installment of the Traveling Stones contains a greater weight of scientific detail about geodes and iron nodules (detail, plausibility). De Quille is also developing the codependence between belief and doubt in his hoaxing. By exposing his hoax in 1879, he simultaneously validated the convictions of those who had originally disbelieved the story and undermined the convictions of the original believers. His tactics of reviving the hoax in 1892 complicate his relationship to these readers enormously. This time he claims that he himself, once a doubter, has been made a believer by the appearance of "testimonials" to the stones' existence. His stated plan to "flush out" the real stones by publicizing fake ones has worked; he has literally made the stones materialize with his rhetoric. Now, De Quille's original believing readers are vindicated, and it is the doubters' turn to be ashamed of their lack of faith in De Quille.

In "The Force of Falsity," the first essay in Umberto Eco's 1998 collection *Serendipities: Language and Lunacy*, Eco discusses historical situations very close to the one De Quille constructed with his "feeler" story about the traveling stones. Eco examines important forgeries and shows how they sometimes led, outside the intention and control of their creators, to serendipitous discoveries. For example, a wishful underestimation of the earth's circumference based on ancient Egyptian geometry by a fifteenth-century geographer motivated Columbus to attempt a westward route to the East Indies, thus leading to the discovery of the West Indies. Eco's argument extends beyond the merely historical, however; in addition to showing how documents could change the course of history, he suggests that intuition and desire, in the absence of empirical fact, are powerful heuristics of discovery. Eco's examples tend to conflate deliberate forgeries with self-delusional theories because his focus is really on how these documents were read and acted upon by others. De Quille's Traveling Stones hoax was a deliberate hoax, not a self-delusion, and his argument that his hoax turned up the "real deal" is equally a fabrication as far as we know. Primarily the hoax served a function outside Eco's field of view—the embarrassment of "outsider" eastern businessmen and scientific entrepreneurs. But De Quille's "feeler" strategy, when viewed from the angle of Eco's analysis of historical forgeries, forges historical authority for De Quille as a literal creator of the West and its scientific phenomena.

In conclusion, the twenty-five year attenuation of the rhetorical game De Quille played with his readers through his Traveling Stones hoax makes two important epistemological arguments: first, that truth is often judged simply as a function of persistence, and second, that readers' ongoing debate over the verity of a story merely serves to cement the authority of its writer as an oracle of natural reality. With the Traveling Stones saga, De Quille developed strategies of testimony that kept his readers ever on the edge of their judgment, all the while cleverly increasing their reliance on him for "information and detail," as Neil Harris put the case for Barnum's authority over his audience.[45] As De Quille's readers constructed their realities, however they chose to do so, they had to go through him and his words. He was a rhetorical magician who made stones literally appear and disappear at whim. It is interesting that eight years after his last installment, stones that appeared to move on their own were reported at the Racetrack Playa in Death Valley, and they are called "traveling stones" by many to this day.[46]

Eyeless Fish (1876)

The Eyeless Fish story was De Quille's final major hoax, not counting the sequels to the Traveling Stones. It appeared as "Mystery of the Savage Sump" on 19 February 1876:

> A most singular discovery was yesterday made in the Savage mine. This is the finding of living fish in the water now flooding both the Savage and Hale and Norcross mines. The fish found were five in number, and were yesterday afternoon

hoisted up the incline in the large iron hoisting tank and dumped into the pump tank at the bottom of the vertical shaft. The fishes are eyeless, and are only about three or four inches in length. They are blood red in color.

The temperature of the water in which they are found is 128 degrees Fahrenheit—almost scalding hot. When the fish were taken out of the hot water in which they were found, and placed in a bucket of cold water, for the purpose of being brought to the surface, they died almost instantly. The cold water at once chilled their life blood.

In appearance these subterranean members of the finny tribe somewhat resemble gold fish. They seem lively and sportive enough while in their native hot water, notwithstanding the fact that they have no eyes nor even the rudiments of eyes. The water by which the mines are flooded broke in at a depth of 2,200 feet in a drift that was getting pushed to the northward in the Savage. It rose in the mine—also in the Hale and Norcross, the two mines being connected—to the height of 400 feet; that is, up to the 1,800 foot level. This would seem to prove that a great subterranean reservoir or lake has been tapped, and from this lake doubtless came the fish hoisted from the mine last evening.

Eyeless fishes are frequently found in the lakes of large caves, but we have never before heard of their existence in either surface or subterranean water the temperature of which was so high as is the water in those mines. The lower workings of the Savage mine are far below the bed of the Carson river, below the bottom of the Washoe lake—below any water running or standing anywhere within a distance of ten miles of the mine.[47]

The fantasy of this hoax ties in with one of De Quille's "quaints," cited by C. C. Goodwin, about a subterranean lake connecting all the Washoe from Lake Tahoe.[48] De Quille consistently worked and reworked ideas that attracted him over time, as evidenced with the Traveling Stones. He had clearly been intrigued by reports of unusual fish for quite some time, because his clippings include an article about ten-headed fish supposedly found in China.[49]

De Quille's own "fish story" ended up being reprinted extensively and attracting high-level scientific attention. De Quille pumped the public enthusiasm for the fish with three and possibly four follow-ups. A New York paper, probably the *Sun*, reprinted the story verbatim.[50] Reactions to the story by local papers were split, and the argument quickly derailed into the issue of water in the mines signaling the running-out of the Comstock lode. The Grass Valley *Union* reprinted the story and reflected, "We regard those fish as evil omen, so to speak. A big cavern full of water will not probably contain much silver ore." The paper went on to carp that the San Francisco merchants who were already refusing to take silver "trade dollars" had better mend their ways before silver production fell off dramatically, and silver became more dear than gold.[51]

The San Francisco *Stock Report* did not like the conclusions the Grass Valley *Union* drew from the fish story. Its writer grumbled that the *Union* was not alone in its naïve assumption-making about the "canard" printed by the *Enterprise*:

> That the story was a palpable "yarn" on its very face to all who understand the conditions of the great mines on their lower levels does not in the least prevent its gaining credit among people who do not understand those conditions, and as the obvious inference is that where there are fish, eyeless or otherwise, there must be water, the story was calculated to injure the mine. A joke is a joke, but such a joke as this becomes serious in its consequences in proportion as it is successful.[52]

De Quille followed up with at least three installments, all of which appear to be from the *Enterprise*. The first appeared the next day and showcased De Quille's famous performance of doubt:

> The local department of the Enterprise of yesterday contained a very nice yarn about fish being found swimming in the water which is now flooding the lower levels of the Savage and Hale & Norcross mines. It is a very Dandequillish story, which, being told on the authority of Col. F. F. Osbiston, Superintendent of the Savage mine, makes us believe it is perfectly true. . . . In fact, the water tastes and smells a little fishy, like the story, and if the fish were a little thicker, it would be merely one extensive chowder.

However, De Quille went on to offer proofs aimed at his readers' expectations of plausibility, through analogy once more (plausibility, analogy):

> Strange as this story may appear to the unscientific, yet it is by no means so unnatural as it seems for even the extreme of cold does not always destroy piscatorial life. We have seen small fish frozen solid in cakes of ice for weeks and when thawed out gradually they all came to life and swam about as lively as ever. . . . An uncle of ours was mate of a New Bedford whaler in the Summer of

> 1848, on the coast of Greenland. One day they found a small whale frozen into the side of a huge iceberg. They cut him out, got a clove hitch about his tail, and left him in the water over night to thaw the frost out of him. In the morning they found him alive and towing the ship to windward at the rate of five knots an hour.

In spite of the connotations of the "whale" with respect to his story, De Quille finishes off his proof with an appeal to eyewitness—his:

> Since writing the above, Mr. James Orndorff, of the Delta Saloon, Virginia City, as shown us some of those fish from the flooded Savage mine, their red color is evidently produced by the oxide of iron, found so plentiful in some portions of the west wall of the Comstock. The flesh is very firm, fins and tail short and compact, and the skin rough and corrugated. They have no scales, and look like a cross between a lobster and a sardine. They can be seen at the Delta saloon.[53]

This item was coupled with an announcement of the exhibition of the fish, "of a variety well enough known to naturalists" at the Delta saloon attended by "hundreds of prominent citizens," which parade De Quille probably surveyed for himself with amused satisfaction over a few beers, as the saloon sits to this day just steps down the boardwalk and across the street from the *Enterprise* offices.[54] De Quille worked all his tricks in those two sequels: *refutatio*, the "Emperor's New Clothes" appeal, analogy, witness, and the use of the sequel to perpetuate the illusion of reality for the hoax, intensify the belief/doubt codependence, and thus cement his authority.

Not content with his success to that point, De Quille provided one and possibly two more sequels to the Eyeless Fish business. One included the textual authority of a forged "letter" from Maurice May of Franktown, Nevada, claiming that at Washoe Lake lived a little gray version of the red eyeless fish; Mr. May surmised that in the journey through the subterranean water tunnels, the fish must have become blind. He also stated his intentions to sue the Comstock for stealing Washoe Lake's water and fish.[55] The last installment, continuing in this tongue-in-cheek vein, may not be De Quille's but certainly sounds like it. It takes issue with the Grass Valley *Union*'s alarmist rhetoric and reassures the reader that the only subterranean sea life they need fear is the "mining shark." "This is a terrible creature, with a stomach and throat extended enough to swallow a city at a gulp. This fish has grown very fat since the Comstock was discovered, and the only redeeming thing in his character is the fact that he prefers his own species for food."[56] The article turns into a comic allegory of the San Francisco and eastern mining interests scamming Nevadans.

De Quille's mature hoaxing tactics in the Eyeless Fish earned not only a wide readership for the hoax but, what is more important, a request from Spencer Baird, curator of the Smithsonian Institution, for a specimen of the fish preserved in alcohol. Thomas Donaldson, Baird's secretary, wrote to De Quille in a letter dated 7 March 1876, "If the statement in the slip enclosed be true, a very important discovery has been made." De Quille's satisfaction with his skillful hoaxing is apparent in the comment he scribbled on the back of the envelope of Donaldson's letter: "A Sold Professor—The 'Eyeless Fish' biz."

Minor Hoaxes and Scientific Reaction

De Quille wrote many other humorous scientific pieces. Loomis counts twelve other "tall tales" of scientific wonders, including another fish story, an article entitled "Ringing Rocks and Singing Stones," and "The Eucalyptus," reporting a new animal of that name found in the Washoe region. Many of these are brief and fall short of the level of scientific plausibility and tone "*au grand serieux*" that characterized his four major hoaxes. De Quille was full of ideas for more of these stories. His notebooks sketch out ideas for "A Natural Incubator—At Steamboat Springs, hatches all kinds of fowls from a humming bird to an Ostrich—I interview the man who has burn marks and [illegible]." De Quille attaches a news clipping to his notebook about an "ossified man" who slowly turned to bone. He notes, "I find similar man hidden in a hut awaiting death. He expects soon his heart will turn to bone—kidneys, liver, one lung gone. He is trying vegetable diet (or some diet containing no lime). Lime in everything. He tries to precipitate it, etc."[57] We will never know if De Quille was going to develop these stories along a humorous line or a more straight-faced line that would have made them good candidates for hoaxes.

Naturally, it is possible that the reason his four most famous hoaxes succeeded where the minor ones failed is due as much or more to reader interests and agendas. The Eyeless Fish certainly hit a sore spot with readers fearful of a future linked to the fate of silver in the West. The Solar Armor hoax arrested reader attention with the conundrum of freezing to death in a desert. The Silver Man, however, attracted almost no attention. I treated it here simply because it was the site where De Quille developed the strategies he would perfect in the successful hoaxes, and it is also the longest and most elaborate of his hoaxes.

I have found evidence of at least four more "serious" hoaxes De Quille wrote, and one of them succeeded in attracting national scientific attention. The underground lake hoax mentioned by Goodwin appears to have been locally successful, but it is hard to track down without further identification. Wells Drury claimed success for De Quille's "quaint" about a windmill that stayed in perpetual motion by using loose sand it hauled up in windy periods as ballast to drive the mill in windless periods; the quaint was supposed endorsed by "an engineering journal," in which a Boston engineer figured out "the exact horsepower" the windmill would generate, but no corroborating evidence for this anecdote is extant. "An Astounding Discovery.—Extraordinary Advance in Science—A Savant Makes an Egg and Hatches Therefrom a Live Bird" appeared in the 19 February 1876 *Enterprise* and claimed to be a reprint from the *Church Union*. This "hoax," however, may have been read as humor rather than as science news due to its outlandish subject (thus violating readers' plausibility expectations) and a few off-key phrasings in the piece like "his darling scheme" and "The Professor was almost wild with delight" (violations of popsci. expectations).

De Quille's Mountain Alligator or Highland Alligator hoax, however, prompted a letter from no less a personage than famed evolutionary paleontologist Edward Drinker Cope. Cope addressed the letter to "Wm. Blackheath" care of the *Enterprise*.

> Sir—
> I see a notice of your "Mountain Alligator" in the Virginia City Enterprise. I do not know how true the statements are, & I write to inquire about the matter. Can you tell me if the length is 7 feet long as stated? Also will you describe the color of the beast?
>
> As a naturalist who has devoted more attention to the reptiles than any other man in this country, I am very curious to give a scientific account of your beast in the Magazine above named [*The American Naturalist*]. Can you find the skeleton & the skull & feet you took from the skin? I can determine exactly what it is if you will send them to me. You may have thrown them away, but I will value them—even if dirty and broken, as I can clean & study them. Of course the skull is most important. Can you send it? I would much like also to see the skin.
>
> Will you give me an account of the locality from whence the specimen was obtained? I will be in Arizona in October. . . .
>
> Yours very truly,
> E. D. Cope

De Quille annotated the envelope, beneath the address, "A Professor who was sold on the 'Highland Alligator.'" The letter is dated 18 September, either 1880 or 1888. A search of the *Territorial Enterprise* for both of those years up to that date revealed no story about a Mountain Alligator or Highland Alligator, but it could very well be that Cope was looking at a paper a year or two older.

Cope may have shared many of the values of Twain's miner readers discussed in chapter 4. Like the local miner population, Cope could not be too picky about which stories he believed if he wanted to beat Othniel Marsh out as the premier fossil hunter of the late nineteenth century. Cope does mitigate his request with "I do not know if the statements are true," but his eagerness to get his hands on the fossil in the second paragraph is almost cloying. An additional dynamic indexed by this letter and the letter from Spencer Baird's secretary is important to our analysis of De Quille's hoaxing. De Quille inscribes on both of these letters the mark of his triumph over the East Coast "professors." He has "sold" the university men on eyeless fish and mountain alligators simply because the professors are outsiders and are therefore not privy to the local, contextual information that would expose the hoaxes.

The use of "sold" is not accidental. Nevada's silver resources were being bought by the government, first to finance the Civil War, and then to finance America's burgeoning foreign trade. This business was being transacted increasingly to the detriment of the local miners and prospectors. De Quille seems to have had a strong sense that his doughty local prospectors were constantly in danger of having their hard work bought out from under them cheaply by eastern commercial interests. This national "yard sale" included the begging, buying, or stealing of western natural wonders such as fossils and geodes by university collectors. Jealous and proud of the progress Nevada had made as a state and as a monument to the pioneering spirit, De Quille got a great deal of satisfaction out of Baird and Cope "buying" a Nevada hoax. That "sale" leveled the playing field a little for the pioneers against eastern money and political power. De Quille once wrote to his sister that he believed Easterners were afraid of the Westerner's "off hand and irreverent way of mentioning men of note and standing."[58] De Quille believed he and other Nevadans possessed, by constructing their destinies on the say-so of no one but themselves, a weapon that was capable of neutralizing the old social institutions and hierarchies that they had moved west to escape but that had followed them anyway. These conclusions based on Cope's and Baird's letters will be addressed more fully in the final section of this chapter.

Summary of Reading Expectations Based on De Quille's Hoaxes

Dan De Quille utilized at least four new strategies in his hoaxes: the "Emperor's New Clothes" appeal, *refutatio*, an eyewitness form of corroboration, and exploiting the codependence of belief/doubt through sequels. Each of these strategies is fundamentally linked to reader expectations. The "Emperor's New Clothes" strategy can be connected with novelty, as it induces a strong desire to "keep up with the Joneses" and appear on top of new developments in science and technology. But it more properly belongs to a level of reader desires that is beyond expectations about ethnoscience and science news, the expectations to which we have restricted this study. "Emperor's New Clothes" appeals tap into a type of social competition fostered by Jacksonian rhetoric about the self-made common American—the struggle to appear independently knowledgeable in all circumstances and thus self-sufficient.

Refutatio is De Quille's strategy for increasing the effectiveness of his argument by constructing a model of his readers' interpretations on-line and adjusting his arguments to play to their highest-ranked expectations. His shifting of his arguments toward plausibility show us that he believed his readers to rank plausibility very highly. The skeptical demands by New York and English editors for further proof of De Quille's claims, the requests by the Smithsonian and Cope for verification, and De Quille's multiplication of eyewitnesses for his local readers support the continued high ranking of corroboration. Finally, it seems likely that many science newsreaders, especially eastern and European readers, were entering a phase where novelty and sensation were being either valued negatively or simply demoted in value because of bad experiences with trusting medicine-show and Barnumesque scientific claims. However, in the case of the Eyeless Fish causing near panic among the Nevada and California papers over the prospect of the mines running out, we see the residue of the reading filters of western miners who still rank novelty higher than plausibility because their unstable futures depend on reacting quickly to new information.

De Quille's performance of the codependence of belief and doubt through his sequels is the area in which we have the most evidence for development and evolution in his hoaxing. Its basic rhetorical success seems to hinge on De Quille's ability to project himself over time as a skeptical authority weighing his options just like the reader. This performance increased the appearance of cognitive similarity between the reader and writer, in Kaufer and Carley's terminology, and thus increased the chances that the reader would be sympathetic to the hoaxer's claims.[59] However, it also made a fundamental, indirect argument for De Quille's authority, namely, "Whichever way you make up your mind, you're still counting on me for your information."

After a consideration of Twain's hoaxing, we determined the following filter of expectations for a western readership predominantly composed of miners and prospectors:

Corroboration, Authority >>
{Novelty, Sensation, Plausibility}>>
{(Medium), Popsci., Internal Coh.}
{Sensation, News, Progression} >>
Comprehension

The top tier reflected the trumping power of reference to authorities outside the physical text for confirmation. For unsigned articles, however, the urgent need to exploit new information (as well as a thirst for sensation on the sparsely populated frontier) tended to win the day over the more quotidian evidence provided by the text in the form of conformity to generic conventions and internal logical consistency. Medium was left in parenthesis because for many readers, it may have been deactivated by the ambiguous reputation of the *Territorial Enterprise*. The bottom tier reflected both Twain's and the contemporary editors' observations that people tended to skim science news, using the familiar structure of the news article as a hypertextual guide for skipping around; this skimming was performed to the detriment of comprehension of details, especially at the end of articles.

After considering De Quille's hoaxing, I cannot make substantive changes to the ranking echelons of the common reading filter. It is tempting to say that eastern/European readers ranked their expectations differently than local miner readers—that miners still ranked sensation and novelty up there with plausibility, whereas more distant urban readers downgraded the first two assumptions and gave more weight to the reputation of the medium. However, the number of responses from each of these two diverse groups of readers is insufficient to warrant creating two separate filters (although intuitively that would be the right step). De Quille certainly was aware he was writing for two different groups of readers, as we will discuss in a moment.

We can, however, see evidence that the definitions of some of the expectations were changing with time. Sensation seems to have been acquiring a permanent negative valence in science news. And corroboration seems to have morphed into something closer to witness for local readers, as they did not rely on textual support for confirmation of a news story, as did more distant readers, but instead on talking to other locals or walking over to the Delta saloon to see if there really were a bucket of red, eyeless fish behind the bar.

De Quille's Hoaxes Build and Defend His Ideal West

The questions remaining about De Quille's hoaxing now are the same that we have answered for each hoaxer: Why did De Quille choose hoaxes to address his readers about science and technology? What message was he trying to get across about those topics? What relationship was he attempting to construct with his readers through the hoaxes? The short answer is that De Quille's hoaxing was overall a constructive activity. Instead of having a particular axe to grind with the way science or technology were being implemented in the West (after all, he was a miner and loved science), he concentrated on using scientific rhetoric to build the West that he wanted to live in and that he wanted to project to the East. He playfully exploited the authority he had earned with his readers through legitimate mining reporting to construct for them and for outsiders a West full of wonders. If, in the process, he caught some important eastern scientists and businessmen in his net, all the better for his project of championing the pioneer as a scrappy folk hero triumphing over the silk-vested eastern fat cat.

This portrait of De Quille's hoaxing revises two previous conceptions of it. C. Grant Loomis's study of De Quille's hoaxes, the most extensive, classifies them as "scientific tall tales" since they are all on topics of science and technology. However, this purely topical approach unfortunately lumps together two different registers of De Quille's scientific journalism as discussed at the opening of this chapter: his humorous "tall tales," and his "quaints" or "sells" (his hoaxes). A tall tale such as "The Boss Rain-Maker,"[60] about a miner who makes rain by shooting the clouds with buckshot, is written in dialect and makes no pretense to be anything other than a humorous story on a quasi-scientific topic. The Silver Man, hoax, however, is written as a high-register science news article. De Quille was clearly engaging his readers in two different games with these two different rhetorics. Lawrence Berkove recognizes this rhetorical difference when he separates out from De Quille's humorous fiction the special category of his "quaints," in which his purpose was "always the same: to gull unwary readers by his matter-of-fact style and copious use of speciously corroborative detail into believing them true."[61]

The distinction is a crucial one to make, for if we obscure it we rob ourselves of an explanation for the resilience of De Quille's reputation as a trustworthy scientific writer in spite of his copious hoaxing. Hoaxing, as discussed with respect to the authority expectations in chapter 3, creates a condition of expert notoriety for the hoaxer; in many instances, this effect provides powerful motivation to hoaxers such as Poe and Twain who wish to create an ethos of countercultural authority. But Poe and Twain's journalism was more amenable to contamination by the waggish reputations they developed through hoaxing. Poe wrote fiction and poetry and essays on those same topics. Mark Twain, too, apart from a short stint as a congressional reporter at the start of his career (which he carefully marked with his given name in order to quarantine it from his developing reputation as a humorist) wrote fiction and satire. They were yarn spinners. It was easier for their readers to reconcile their hoaxing with their overall literary endeavors. De Quille, however, evinced a rather sharp divide in his journalism between his technical writing on the one hand and his tall tales and hoaxes on the other. The contemporary commentary shows De Quille's readers cognizant of this Great Divide. Wells Drury, for example, writes the following about the relation of De Quille's hoaxing to his technical writing: "When the newspapers of the coast took Dan to task for his trifling, Dan only laughed and resolved never to do it again, but the next time that items were scarce he was tempted and fell from grace. . . . These diversions, of course, were only occasional and desultory. In his regular work Dan was a model of method and accuracy. This made his hoaxes all the more dangerous."[62]

Because De Quille had two distinct journalistic modes, his readers were able to bracket off his technical writing and reputation from infection by the notoriety that hoaxing usually brings. It is interesting, however, that that interpretive barrier was permeable from the technical side, as Drury notes above: De Quille's reputation as a scientific expert continued to lend at least initial credibility to his hoaxes, written as they were to sound like serious science articles. Local readers had access to enough ancillary evidence to eventually sort out De Quille's quaints from his technical pieces. East Coast readers likely did not. With both readerships, however, De Quille's formidable technical expertise appears to have insulated his journalistic ethos from notoriety.

De Quille scholars have made another, more serious, error in appraising his hoaxing. Much like the critics who claimed Twain had no social project, Lawrence Berkove argues that De Quille's hoaxes had "no ulterior purpose beyond entertainment" (20). As we have seen throughout this book, hoaxing is serious business, altering the very fabric of what readers consider real. The editors of the San Francisco *Stock Report* cut to the core of the issue when they criticized the Eyeless Fish for starting a dangerous rumor about the future of silver in the West: "A joke is a joke, but such a joke as this becomes serious in its consequences in proportion as it is successful." A hoax becomes epistemologically and politically serious when many people start believing it. That De Quille was serious about people believing his hoaxes is evident in the copious sequels he wrote to the Traveling Stones and Solar Armor hoaxes in particular. He persisted in defending his hoaxes for as long as thirty years.

Why would De Quille work so tirelessly to secure belief in his hoaxes? Most of the locals that read his paper could verify within hours or days that De Quille's "quaints" were nonsense, just by asking the people referenced in them. Some answers to this conundrum become clearer when we consider De Quille's second readership, his East Coast readers, along with his central social and political passions: state building and the free silver movement.

State-Building and the Prospector as Folk Hero

Lawrence Berkove justly describes De Quille as a major western writer who "was shaped by the West even as he helped create it."[63] De Quille was invested in Nevada. After all, he chose to spend nearly forty years of his life there when he had a wife and three children back east in Iowa. His papers contain envelopes in which he collected clippings pertaining to "Nevada matters,"[64] and he followed and wrote on the politics of statehood. He owned a complete copy of the laws passed by the Nevada legislature during the 1863–1864 session, right before Nevada achieved statehood.[65] He literally defined "Nevada" for the *Encyclopedia Britannica* in 1884.[66]

De Quille's hoaxes are not about events in Europe or on the moon, as Poe's were. They are all about Nevada locations, phenomena, and people. De Quille was a correspondent for eastern papers and knew they clipped his stories. He knew that hoaxes created new realities for readers, both western and eastern. The hoaxes, along with De Quille's other stories, served to create a larger-than-life legend of Nevada as a scientific wonderland and Nevadans as stout-hearted individualists who did not shy away from the most bizarre of discoveries. As De Quille wrote of the men who supposedly discovered the silver man, "Had the finders been any other than California or Washoe miners, there would have been a jolly stampede and some frantic climbing" to retreat from the terrifying sight.[67] De Quille did not insert people in his hoaxes, as Twain did, to mock them. The prospectors, miners, inventors, and doctors who staff De Quille's hoaxes are America's new folk heroes—ingenious, fearless, and occasionally tragic, as in the case of Jonathan Newhouse and his solar armor.

One might justly argue that the self-revealing character of De Quille's hoaxes would damage the grand reputation he was trying to construct for the West and westerners with an eastern readership. In fact, the track record of revelation for De Quille's hoaxes is extremely interesting, as he only revealed one in print, the Traveling Stones hoax. His other hoaxes, much like Poe's and Twain's, contained clues and witnesses that locals could use to debunk the stories. The explicit revelation of the Traveling Stones, as we saw in the analysis of that hoax, was not so much a revelation as a dramatic prelude to De Quille's final installment in the hoax, his own conversion to a believer in the existence of the stones. Cleverly, he claims to have believed from the beginning that they were always out there and that his story merely goaded his readers on to find them. Revelation, therefore, was actually a constructive strategy for De Quille, allowing him to capitalize on the codependence of belief/doubt to assert his authority over the reader who could not independently test his quaints. This reader was the eastern reader, the main target of De Quille's hoaxes.

Free Silver and the Defense of the West

Nevada became a state in 1864, and De Quille lived there for nearly thirty years after that. Lawrence Berkove writes of his later years, "He had outlived the bonanza times but was not content to silently carry to the grave his love of the pioneer and prospector ethic."[68] For this reason, Berkove claims, De Quille wrote *Dives and Lazarus*, his longest fiction work, which remained unpublished until nearly a hundred years after his death. Through the allegory of a poor prospector who goes to heaven and a rich gold trader who goes to hell, the book contains some heavy propaganda on an issue that engaged De Quille's passions—the free silver movement. When viewed through the lens of this political issue, De Quille's hoaxes appear as a means to symbolically redress the debts that he felt eastern governmental and commercial interests owed the Nevada prospector.

The free silver movement was a reaction to eastern exploitation of western resources. Nevada was made a territory in 1861 by Lincoln so that its burgeoning silver yields would serve the Union and not the Confederacy during the Civil War. The silver financed a great deal of the rest of that war as well as foreign trade afterword, as noted above with respect to the "trade dollar" debate in the Eyeless Fish hoax. However, in spite of the boon the vast silver resources of Nevada had provided the nation, the United States stuck to the gold standard after the Civil War, a policy that increasingly hurt Nevada during the series of depressions that wracked the country from the 1870s until World War I. Silver languished at various unpredictable fractions of the price of gold, making paying for debts and basic necessities with silver dollars nearly impossible. Proponents of "free silver" wanted the U.S. mint to allow the stamping of as many silver dollars as there was free silver in Nevada, at a constant rate of sixteen silver dollars to one gold dollar. This would have helped miners and small businessmen in the West recover somewhat from the depression. But the U.S. government refused, and miners believed it was because big businessmen and bankers, who would suffer if the debts they held were easier to pay back, had undue influence with the government.[69]

In addition to the allegory of *Dives and Lazarus*, De Quille wrote vehemently in favor of free silver. In his articles he polarized everyone into "the millionaires versus the masses, plutocracy versus democracy, remorseless Shylocks versus the 'race of hardy frontiersmen,' evil versus good.'"[70] It was in this phase of his life that De Quille, who had been multiculturally educated and tolerant in his fiction and nonfiction to this point, suddenly wrote bitterly against Jews and the Chinese, whom he believed were conspiring with the gold-standard supporters against the miners.

This passionate defense of the miner against the East Coast political and business interests dovetails with De Quille's hoaxing practices. As mentioned with respect to the Eyeless Fish and Mountain Alligator hoaxes above, De Quille enjoyed "selling" big East Coast professors on stories that his uneducated miner friends at the saloon would never take seriously for more than a day or two. The professors were outsiders, Nevadans were insiders, and the hoaxes dramatized that divide. The same dynamic held in the Traveling Stones revelation of 1879. De Quille claimed he only revealed the hoax because he was tired of being pestered by requests for the stones from P. T. Barnum, German scientists, and eastern businessmen. Perhaps his motivation was rather a symbolic victory over these prestigious figures via public humiliation. Remember that he rounds out his revelation by sniping, "If this candid confession shall carry a pang to the heart of any true believer we shall be glad of it, as the true believers have panged it to us, right and left, quite long enough."[71]

Conclusion

De Quille's hoaxing is not an indirect criticism of the role of science in American life, as was Twain's and Poe's. De Quille was a lay scientist and wrote as enthusiastically about the progress of science and technology in the West as anyone. The target of his hoaxes was the East Coast commercial appropriation of what western miners and pioneers broke their backs for—a problem that he also worked out in detail in his free silver journalism and in *Dives and Lazarus*. The hoaxes were De Quille's private, and public, vendetta against the powerful eastern interests victimizing De Quille's friends and neighbors.

In the end the legacy of De Quille's hoaxing moves beyond the valiant but losing fight he waged against the East and the gold standard. His hoaxing practices utilize the foundational strategies developed by Poe and adapt them to a popular science rhetoric that demonstrated through its matter-of-fact language the control that American science had extended over nature in the intervening thirty years. In addition, De Quille's hoaxes clearly demonstrate their inheritance of Twain's conception of the hoax as local political activism. But De Quille is also an innovator. He develops rhetorical strategies of hoaxing—like "Emperor's New Clothes" appeals, forged testimonies, and sequels—that specifically exploit the power of print to build reality over time. His legacy is the idea of the West as a perpetual frontier where the possible and impossible regularly change places through the translation of scientific rhetoric, as a liminal realm that divulges its mysteries only to those who go there and write about it. With his hoaxes De Quille—just as significantly as other, more canonical western writers such as Twain and Bret Harte—conditioned his readers to believe that the West was a region of American experience where words, in fact, made the world.

Chapter Six

The Mechanics of Hoaxing

While Dan De Quille was writing his hoaxes, John Muir was wandering through the High Sierra taking an overview of human activities in the region. In "Nevada's Dead Towns" he makes an uncharacteristically harsh judgment on the efforts of De Quille's mining colleagues:

> Nevada is one of the very youngest and wildest of the States; nevertheless, it is already strewn with ruins that seem as gray and silent and time-worn as if the civilization to which they belonged had perished centuries ago. . . . Wander where you may throughout the length and breadth of this mountain-barred wilderness, you everywhere come upon these dead mining towns, with their tall chimney-stacks, standing forlorn amid broken walls and furnaces, and machinery half buried in sand, the very names of many of them already forgotten amid the excitements of later discoveries, and now known only through tradition—tradition ten years old.[1]

Muir goes on to contrast the waste with the mining towns of California, which he compares to the gracefully aging ruins of Europe whose "picturesque towers and arches seem to be kindly adopted by nature. . . . They have served their time, and like the weather-beaten mountains are wasting harmoniously." But the dead Nevada towns "do not represent any good accomplishment, and have no right to be. They are monuments of fraud and ignorance—sins against science." Muir concludes his diatribe with a benediction that resurrects the phantom of the miners' stubborn attachment to their dreams: "But, after all, effort, however misapplied, is better than stagnation. Better toil blindly, beating every stone in turn for grains of gold, whether they contain any or not, than lie down in apathetic decay."[2]

Muir's reaction to the mining towns is complex and telling. He understands gold fever and rates it above an adventureless life. But he forges a unique alliance among beauty, science, and existential rights and then says that the "machinery half buried in sand" of Nevada's mining towns violates this compact. Why are Nevada's abandoned camps "sins against science"? Muir seems to be arguing that their cardinal sin was their foundation on "fraud and ignorance" rather than systematic proofs of their prospects. The California towns are beautiful because they are monuments to fruitful labor, for however short a span; the broken Nevada works are desecrations because they memorialize nothing but rumors, wasted efforts, and wasted lives. Their spectacle of failure taints legitimate scientific ventures.

Muir was talking about ruined mining works, but he could have been talking about the scientific media hoaxes considered in this study. Their machinery was awesome, convincing by virtue of its illusion of smooth operation, but in the end there was nothing behind the rumors that had constructed it. Nevertheless, its spectacular failure did not prevent readers from rushing after the next rumor, and the next. And Muir celebrates the backward virtue in this rushing to believe: it is better than believing nothing.

The connection between Muir's dramatic portrait of the broken mining machinery and the hoaxes as we look back at them "half buried in sand" is more than an artistic convenience, as I will argue in a moment. There is contemporary support for an essential connection between machines and hoaxes. But before I embark on answering the last question posed by this project—What is a hoax?—I will summarize the answers to the first two pragmatic questions that began my research: How did the hoaxes work? And what were the hoaxers trying to accomplish?

How Did the Hoaxes Work?

The scientific media hoaxes under examination in this book worked by performing reader expectations about ethnoscience and science news. This performance invited belief and then effected a public revelation that opened the readers' strongest assumptions up for criticism. Modeling this interpretive process required a methodology that could cope with the entire communicative loop of the hoax—author, medium, audience message—or the "symbolic action" the hoax enacts in its community, to use Karlyn Kohrs Campbell's terminology.[3] The OT-based method chosen for this analysis meets these criteria because not only does it model the interaction/competition of readers' expectations during their decisions about the truth of a hoax, but it also models the guesses of the hoaxer as he strove to camouflage his project under features that readers would expect to see in a "real" science news item.[4]

That being said, we must examine two consequences of approaching hoaxing as a rhetorical game in guessing and reperforming reader expectations: first, authorial intention could not be ignored, as it turned out to be a crucial feature of the hoax as a social project; and, second, the reception of hoaxes at different times and places (kairoi) in nineteenth-century America was represented as (evolving) filters of common reader expectations.

What Were The Hoaxers Trying to Accomplish?

Each author had a slightly different social project in mind with his hoaxing. However, a general answer to the question What were the hoaxers trying to do? can be framed, and this generalization is enabled only by a view of the hoaxes from the point of view of reader expectations. Scientific media hoaxes serve to surface readers' strongest assumptions about science news and ethnoscience. These top-ranked expectations are then available for criticism or revision. All of the hoaxers in this study keyed their projects of indirect social criticism to readers' top-ranked assumptions about science such as authority, novelty, and sensation.

The Social Projects and Intentions of the Hoaxers

Poe's hoaxing was groundbreaking. He and Richard Adams Locke innovated the genre just as Poe innovated the American genres of the short story, detective fiction, and science fiction. Poe's hoaxes took advantage of current "fads" in popular technology and science, especially those of European provenance. He used these sexy topics as vehicles to launch an indirect assault on a reading culture that assigned authority to Baconian scientists rather than artists and that downgraded knowledge of scientific detail and internal coherence in favor of sensation. He considered the revelation of his hoaxes as a social triumph for artists, reasserting their authority not only over the language of science but also over the construction of public truth. Simultaneously, Poe exploited the double audience that hoaxes construct to materialize for himself a community of like-minded readers who shared his preference for imaginative scientific epistemologies such as the one outlined in *Eureka.*

Twain's hoaxing inherited many characteristics of Poe's practices, especially the criticism of sensation and the undercutting of the authority of paleontologists. While Poe's hoaxes played on the exotic marginality of Europe or pseudoscientific phenomena, Twain's hoaxes were rooted in the West and dealt with local issues. Both hoaxers thus exploited frontiers, lines where readers' knowledge became fuzzy or unstable, but they did so for different purposes. Poe had a personal scientific agenda to push. Twain was disturbed first by scientists' cooption of the right to write the West for Americans and later by the bending of technological progress in America toward imperialistic ends. His hoaxes produced laughter that was an affirmation of self-determination in the face of many American institutions—science, organized religion, Victorian morality, the military—that were vying for authority over the individual. However,

Twain's hoaxes also served to reestablish his authority over his readers as the only trustworthy oracle willing to unmask others' social hoaxes and reveal the real West, the real America.

De Quille's hoaxing was a celebration of the ability of the scientific imagination to create American reality. He continued Twain's local practices, but his personal rhetoric of hoaxing was more complex than either Twain's or Poe's. Manifesting a finely tuned awareness of the needs and expectations of his readers, De Quille actually adjusted his argumentation within his hoax to anticipate his readers' interpretations using a strategy of *refutatio*, and he employed sophisticated rhetorical manipulations of his readers' psychological needs for witness and for insider status in order to secure belief in his hoaxes. His target was also slightly more focused than Twain's or Poe's. Instead of railing against the authority of science or industry, De Quille's hoaxes targeted the authority of eastern government agencies and universities seeking to profit from the risk and labor of western pioneers. De Quille exploited the insider/outsider dynamic of hoaxing inherited from the tall tale in order to defend his ideal West from eastern exploitation.

In addition to these social projects, two of our authors had personal axes to grind with their hoaxes. Poe used "Von Kempelen" to mock George Eveleth, an irritating groupie; Twain famously aimed The Petrified Man at Judge Sewall.[5] Since the hoaxes make public these very personal goals, it is impossible to avoid the issue of author intentionality in hoaxing, no matter how vexed a question it may be.

The easiest formulation of the problem is this: If an author intends a hoax, but no one believes it, is it a hoax? The answer is almost certainly no, and the term *failed hoax* is probably the most felicitous for this situation. The more troubling formulation of the author-intentionality problem in hoaxing remains as follows: if an author did not intend his/her story to be taken for a serious witness of a real scientific event, but it was received as such by a readership (not just one or two people), was it a hoax? My gut reaction, which I suspect is shared by many, is no. The folk use of "hoax" bears negative connotations that felicitate its pairing with verbs such as *perpetrate* and *foist*, which require a villain as an agent. However, at least two complications to this claim have emerged in the course of this book.

One of these complications was the reception of Poe's M. Valdemar. There is no extant evidence that Poe intended the story to be taken literally; however, once it became clear to him that he had secured the belief of readers from Europe and America, he began referring to the story as a "hoax" to pump up its celebrity. This seems to indicate that author intentionality is not simply a problem for the framing of the hoax but also gets interleaved with reader responses. Thus, M. Valdemar is a hoax because Poe chose to own it as such once readers started believing it.

A second problem of intention arose with Twain's hoaxing, where he claimed that The Petrified Man and Empire City Massacre were not hoaxes but satires, and he spent a great deal of time justifying this claim by pointing to clues he had planted in the stories, which his readers had missed. While Twain's claims are not consonant with his actions, which were calculated to perpetuate and repeat these "misunderstandings," this scenario raises the crucial role of reader reception in the hoaxing event. If many readers believe something to be true that is later revealed to be false, irrespective of author intentions, the word *hoax* can still be applied. For example, "War of the Worlds" is almost always referred to as a hoax in spite of the fact that the original broadcast contained announcements that it was fiction; the massive public panic that ensued has ensured its historical status as a hoax.

What seems crucial is that in a hoaxing event, the fooled readers or listeners hold someone perceived as the author accountable for their embarrassment or discomfort. In the case of "War of the Worlds," angry responses from listeners forced Orson Welles to issue a public apology for the "misunderstanding," even though his broadcast had included disclaimers. Therefore, this is the answer to the intentionality problem that I offer for now: if an audience publicly constructs itself as deliberately deceived by an author or agency through a news medium, the event counts for historical purposes as a hoax, irrespective of the stated intentions of the author. This definition rules out a case where one or two readers with active imaginations or serious psychoses believe a piece of science fiction is a news report. This definition also correctly assigns cases such as John Symmes's "Symzonia," in which the author genuinely believed in and lobbied for the report he issued of subterranean passages leading to the earth's core.[6] This was not a hoax because even if a readership believed the story, so did its author. There was no "malice and aforethought," no discrepancy between the knowledge states of the author and readership that could be publicly demonstrated and decried.

This definition of author intentionality in hoaxing actually makes a striking argument for the function of reader response in constituting authorial agency—and in levying the historical judgment of "hoax" itself against these rhetorical exchanges. Reader judgments formed the core of this wook. From readers' reactions to the hoaxes, I reconstructed expectations associated with science and science news that repeatedly served as "sticking points" in the debates over the truth of the hoaxes. These expectations were organized according to their relative power in the interpretive process into ranked sets of expectations that readerships held in common. By comparing these common reading filters across time and between readerships, we can get some perspective on how the reading of science news changed in America from 1835 to 1880.

Reception Figured as Filters of Reading Expectations

The result of the application of this method is an evolving portrait of reader priorities spanning the careers of these three authors. Overall, four general trends are visible upon inspection of table 6.1, which compares the filters of expectations gleaned from the contemporary reactions to each of the three author's hoaxes.

A few reminders about the structure of the filters are in order. The filters are linear lists of reader expectations in order, left to right, of decreasing strength in determining decisions about the truth of the hoaxes. No single reader's interpretation of a hoax is likely to use all of the expectations in the filter, as readers tend to focus on two or three competing expectations, at the most. The common reading filter is an abstraction that synthesizes the ranking information from many different readers' decisions. The crucial levels of rank in the filter are indicated by ">>"; if there is more than one expectation in a level, they are bracketed together in order to indicate their equality of strength and lack of competition with each other in the interpretive process.

A few general trends are visible from comparing the filters. The function of corroboration and the reputation of the author or information source (authority), whether positive or negative, persisted as a powerful determiner of reliability. The reputation of the medium (medium) became a weaker constraint on decisions about truth in western journalism, which was very new and still involved with folk practices of tall-tale telling and practical joking. The placing of medium in parenthesis indicates its optionality. Some readers, like rival editors, used the negative reputation of the *Enterprise* as an important factor in their decisions about potential hoax stories. Other readers, accustomed to the unreliability of the *Enterprise*, did not use medium in their decisions at all; they just used their own "common sense" and the textual, factual expectations such as plausibility, popsci., and internal coherence. Finally, information junkies such as miners and collec-

Table 6.1 Comparison of Filters of Reading Expectations from the Project

Expectations of Poe's New York Newsreaders from 1835–1849
{Corroboration, Medium, Authority}>> {Novelty, Sensation, Plausibility}>> {Popsci., Foreign, Internal Coh.}
Expectations of Western Newsreaders 1862–1880
1. {Corroboration, Authority} >> {Novelty, Sensation, Plausibility}>> {(Medium), Popsci., Int. Coh.} 2. {Sensation, News, Progression}>> Comprehension

tors such as Cope and Baird had to weigh their desire for a scoop (novelty) against the fishy reputation of the paper, so medium would still be active for them at a lower level. Developing separate common reading filters for these different audiences is desirable but impossible given the scarcity of archival responses to these western hoaxes, so the filter represents the reading habits of the majority readership—miners.

The reactions to De Quille's hoaxes in the 1870s, particularly those from the reprinting papers, revealed that novelty and sensation were not considered as reliable indicators of scientific truth as they were a few decades earlier. Eastern readers, especially, tended to evince suspicion without corroboration or additional evidence of plausibility. A final observation is that for local readers of the western hoaxes, corroboration really was defined something more like eyewitness. In the sparsely populated and documented territories, readers used other readers to corroborate scientific claims.

A few other trends are not immediately apparent but still important. The popular science article (popsci.) changed over time to conform more to the format of a regular news article, front-loading the who, what, when, where, and why and foregoing the mystery opening. This development goes hand-in-hand with the suspicion of sensation appearing in the reader reactions to the postbellum hoaxes. The popular science article also began to place background information at the end of the story instead of at the beginning, thus favoring a journalistic rather than a strictly narrative structure. The disappearance of foreign from the expectations of the later western newsreaders reflects Twain's and De Quille's adjustment in their hoaxing to a local epistemology based on lay eyewitness.

Beginning with the 1862 hoaxes, I introduced a second tier of expectations that reflected Twain's and editors' insights into science newsreading psychology. This delayed introduction does not imply that Poe's readers never skimmed to get to the good parts. However, by the 1860s the generic profile we recognize today in news stories was basically set, enabling readers to make use of skimming heuristics. Not visible in this table is the expectation entertainment, which deactivates all decisions about truth and leads to reading with suspended belief, similar to fiction reading. This of course was the stance of some readers who enjoyed the play of the hoaxes. Many readers, however, judging from their strong reactions, took the hoaxes as quite serious games with their realities, games with consequences.

The final and major consequence of a rhetorical redefinition of hoaxing must now be addressed. If a hoax is not a certain kind of text, what is it? Exactly what kind of "rhetorical exchange" is a hoax, and what "symbolic action" did it accomplish in its reading community? Though these questions have been answered on a very local scale for each author, this project—definitional as it is—must provide at least a limited historical generalization. A forty-five-year span of reading and cultural development in America (1835 to

1880) initially seems too much to comprehend with any one coherent, helpful statement. However, if we consider some similarities in the hoaxers' practices over this time, a productive pattern emerges.

A telling similarity in the hoaxers' practices is their absorption with mechanics. Two of Poe's four hoaxes construct machines. Many of his other writings, especially "Maelzel's Chess-Player" and "The Gold-Bug," perform a strong correspondence between the construction of discourse and the engineering of machinery. Twain's hoaxes are not about machines, but much of his other "scientific" fiction—particularly *A Connecticut Yankee in King Arthur's Court*, *An American Claimant*, and *3,000 Years among the Microbes*—are actually about technology and mechanics. In addition, his partnership in the Paige typesetter, his inventions, and his installation of telephones and other gadgets in his own home are just a few instances of a well-documented lifelong fascination with machines. De Quille wrote extensively on machines in his technical journalism, and the Solar Armor hoax is explicitly mechanical.

These preoccupations with creation, authority, social control, and mechanics suggest that the rhetoric of the nineteenth century scientific media hoax may have operated by analogy to the machines that surrounded and fascinated Americans during this time. This conclusion receives corroboration from the hoax's conditions of production, from contemporary commentary connecting textuality and mechanism, and from current historical analyses of machines and culture in nineteenth-century America. In short, the answer to the final question of this book—What is a hoax?—turns out to be that the hoax is a machine.

The Hoax as a Machine

As we have already seen in comparing the hoax to the tall tale, the hoax contains elements that are strikingly industrial when compared to its oral predecessors. The hoax relied on the following machine-age institutions: the communicative distance of print technologies, a mass distribution network, and an industrial mode of authorship that encouraged anonymity as a means of effacing the individual and strengthening the perception of the institutional. Examining the conditions of production of the first media hoaxes reveals that the hoax is indeed a machine-age genre. Most early hoaxes or media satires in Britain, such as *Gulliver's Travels*, *Robinson Crusoe*, and "A Modest Proposal," all date from a time after the Industrial Revolution. In fact, the *Oxford English Dictionary*, second edition, claims the earliest the word *hoax* appears as either a noun or a verb in print is in 1796 in *Grose's Dictionary*; most other usages are midnineteenth century.

Miles Orvell argues in *The Real Thing* that two primary conditions of the Industrial Revolution made fakery possible if not inevitable: first, machines cheaply produced thousands of near-perfect copies of goods, riveting the value of mechanism and fascimile deeply in the public consciousness; second, a booming industrial economy replaced transactions with trusted individuals with repeated transactions with strangers, thus shifting public trust from personal ethos to general templates or schemata for transactions, a focus on form rather than content.[7] Hoaxes were a special rhetorical mechanism for exploiting public trust in form and facsimile in order to display its instability. As such, they were attractive to any writer who wished to draw the critical public eye to key "cogs" in the industrial workings of America such as businesses, the government, and professionalized science.

Comparing discourses to machines became a reflex in the thought of some of America's most prominent philosophers and writers. One of the most famous instances of this connection is discussed by Leo Marx in *The Machine in the Garden*. Ralph Waldo Emerson said by just looking at the workings of a steam engine, he could read as though from a text the industrial progress the machine was engaged to produce; through its gears and pistons, it both announced and interpreted itself as a messenger of progress for the viewer.[8] Henry Adams in the "Dynamo and the Virgin" chapter of *The Education of Henry Adams* figures the social discourses of his era as an enormous, sublime engine spinning almost out of control. Moreover, Mark Twain treats his composing abilities as mechanical potential when he writes his brother Orion, "for the talent is a mighty engine when supplied with the steam of education—which I have not got, & so its pistons & cylinders & shafts move feebly & for a holiday show are useless for any good purpose."[9] It is clear that mechanism, especially the mechanics of the steam presses that produced texts, became a metonym for the production of discourse.

This mechanical metonymy brought with it a fear of runaway effects that could outpace the intentions of the creator of the discourse or machine. Emerson recorded these fears applied to technology and imperialism in his "Ode, Inscribed to William H. Channing" when he wrote, "Things are in the saddle and ride mankind."[10] Henry Adams expressed very similar fears in a letter in 1862: "Man has mounted science, and is now run away with. I firmly believe that before many centuries more, science will be the master of man."[11] These fears of snowballing effects, transferred to rhetoric, become worries about mass production and the decline of quality in literature. Poe himself, in his "broad-axe" criticism, repeatedly condemned the American publishing industry for literally manufacturing "a pseudo-public-opinion by wholesale," aimed at puffing American literature for strictly commercial purposes.[12] He argued this "puffery" was destroying the very literature it sought to "elevate."[13]

These connections between texts and machines were more than just abstractions. They became ingrained at a very basic level of ontology, as readers came increasingly to associate reading and writing with printing, and literature with printed texts. Recently, scholars have discovered even deeper influences between the rhetoric of science texts and the technologies that both produced and justified them. Elizabeth Tebeaux, in her study of Renaissance technical manuals, found that authors often established the utility of these manuals by titling them with the names of contemporary technologies that readers knew and trusted on a daily basis. So a treatise on military maneuvers was titled *The Military Garden*, and two healing manuals were titled *The Castle of Helth* and *The Myrour or Glasse of helth*.[14] These metaphors argued that the text was not just a text, but actually a technology or mechanism for improving the quality of life.

Likewise, the nineteenth-century writers who compared discourses to machines were not invoking an abstraction but were constructing a concrete metonymy to a very particular sort of machinery familiar to their readers: "gears-and-girders" technology, in Cecilia Tichi's terminology. In *Shifting Gears* Tichi defines gears-and-girders technology as machines that visibly transform energy, for example, coal to steam, and that encourage their viewers to imagine themselves as coengineers by laying the structure and workings of the mechanism bare to the novice eye, fostering a sort of "Oh, so that's how it works," gestalt experience.[15] A nineteenth-century scientific media hoax operated in a similar fashion: it transformed readers' assumptions about science into an embarrassed awareness of the instability of those assumptions, and it did this precisely by revealing the structure of those assumptions to the reader. Through this process, the hoax made its readers coengineers; it implicated them in constructing the problem (professionalized science taking over American society) that drove the hoax in the first place.

The ability of the hoax to stand in for a social problem and transfer agency for that problem to its reader/viewer makes it a "hybrid" in Bruno Latour's theory of the relation of science to society as outlined in *We Have Never Been Modern*. According to Latour, all technologies, including popular science articles, instantiate the connectivity between Nature and Society, a dichotomy that the project of modernity has futilely tried to create and maintain through analytical criticism. Latour argues that hybrids are the central irony of a modern world view, because the harder we try to segregate the human and nonhuman elements of our world, the more connectivity we create between the two in the form of technologies to do our science, economic and political alliances to regulate our societies, and texts to clarify our epistemological positions.[16] Technologies, economies, polities, and texts are all hybrids. The popular science article is an archetypal hybrid, as it is a text written by humans that nevertheless conveys, supposedly transparently, a nonhuman or transcendent truth about the world. A scientific hoax, then, can be viewed as a sort of metahybrid. It is a technology whose function is to call attention to the hybridity of the scientific article; it accomplishes this function by forcing readers to confront their dependence on science news, to acknowledge the ways in which hybrids (science news articles) are substituting themselves for social understanding of and judgment about the natural world.

A mechanical model of hoaxing is therefore both immanent from the contemporary culture of Poe, Twain, and De Quille and corroborated by recent analyses of the relationships between machines and people in the modern era. As with any other model, it carries consequences for the hoaxes studied in this book. First, the idea of the hoax as a machine implies that the hoaxes we have examined were engineered with specific input and functions in mind. Indeed, this is the case, as the hoaxers gauged their readers' expectations and then constructed—via word choice, format, and argumentation—a mechanism that satisfied these expectations and produced belief as an outcome. By planting clues or by exposing their machinations extratextually, the authors were able to use the machine of the hoax itself as a lesson. Readers were able to look back at the text/machine and see exactly the processes by which their assumptions were exploited to secure their belief. Expectations were both the motivating principle for the construction of the hoax and its fuel.

The notion of "fuel" surfaces another point of fit between the rhetorical approach taken toward hoaxing in this work and mechanics: both have to be in motion to work. Considered alone and inoperative, machines and hoaxes are mere artifacts of axles, cogs, or words. They have to be in action to be themselves, since both machines and hoaxes are the sum of their functions. This fact helps illustrate again why a hoax is no longer a hoax once it is removed by time or space from its original publication context. Both a machine such as an old combine and a hoax such as M. Valdemar lose their significance when viewed in a state of inactivity and removed from the contexts of their original operation—say in a junkyard or in a science fiction anthology.

Finally, the notion of a hoax as a machine helps illuminate the experimental "tinkering" that all three authors appeared to do with their hoaxing over time. Each writer learned from a certain hoaxing experience what had worked and what did not, and after this analysis adjusted his rhetoric to produce a more successful result the next time. Poe recorded a great deal of this process for us in his writings about Locke's Moon Hoax and Hans Phaall; certainly he produced a much more successful hoax the next time around with the Balloon-Hoax. Twain provided us with a negative portrait of this tinkering process, as he dramatized all of the reasons his "satires" failed and yet continued to persist in those rhetorical practices that coerced belief from readers. De Quille did not leave overt commentary on his revising of his hoaxing practices, but as we have seen, he responded to the success of the Solar Armor sequel by constructing numerous sequels for his remaining hoaxes, thus solidifying his authority as an expert witness to the West while making an implicit argument for the truth of his hoaxes via their persistence through time.

Viewing the scientific media hoax as a rhetorical and psychological machine fits well with the evidence from this book and argues for a feedback loop of influence between salient technologies in a reading culture and the structuring of rhetorical exchanges within that culture. If, as Elizabeth Eisenstein and Walter Ong argue, print technology has affected (but not determined) readers' cognitive organization and function over the last four hundred years, and if as Kathleen Welch argues, the Internet has affected the design of books, documents, and arguments in the classroom, it seems reasonable to suggest that a salient technology such as the gears-and-girders-type machine could catalyze the development of a new rhetorical genre. Being tied to a particular type of technology keeps the hoax/machine correspondence limited locally and temporally, so these claims cannot be generalized to all times and places where hoaxing occurs. I will argue shortly, however, with respect to the recent Sokal hoax, that the mechanics of the scientific media hoax have evolved over the last century to mimic information technology instead of gears-and-girders technology.

Tuning Up: The Hoax Then and Now

The project method exposed a rhetorical mechanics via which writers of hoaxes used reader expectations of science and science news to drive systems of illusion and revelation that created opportunities for indirect social criticism. Reader expectations represent cognitive, psychological, and cultural factors influencing the reading experience; thus, the method continues to explode inherited romantic conceptions of reading as a solitary and individual act between a mind and a text. In this study of hoaxing we have seen that a hoax is not a person being tricked by a text but instead a complex and coordinated social project in which current events, culturally inherited ideas, celebrities, learned conventions of reading, personal agendas, media, and the projected assumptions of other readers all shape public decisions about what counts as the truth.

Hoaxing both reifies and responds to lack: gaps and inequities in status, education, and group membership. These gaps did not vanish with Victorian science; they have merely shifted in size and orientation as the arts and sciences have endured the two World Wars, the Cold War, and most recently, the science wars. In many ways, Poe and Twain lost the fight to keep professional scientists from becoming America's new oracles. Barry Barnes claims that science is our new metaphor for comprehending both our physical and social realities. "For us," he writes in his study *Scientific Knowledge and Sociological Theory*, "natural order is a model for understanding social order."[17]

Yet even within this new order, hoaxing remains a first-rate strategy of comeuppance, a means of redressing power imbalances between the public images of the arts and science through the mass media. Because the social stakes have changed since the nineteenth-century hoaxes, the mechanics of the hoax have morphed to reflect current technologies and address current power struggles. The recent Sokal hoax is the perfect site to examine the evolution of scientific media hoaxing, since it is in many cases the inverse of the hoaxes we have studied—a scientist hoaxing literary critics on the grounds that they are trying to recapture the right to determine what counts as truth (or the very status of truth itself) for the American public.

Conclusion: The Sokal Hoax

In 1996 Alan Sokal, a particle physicist at NYU, submitted an article to the prestigious cultural studies journal *Social Text* entitled "Transgressing the Boundaries: Toward a Transformative Hermeneutics of Quantum Gravity." Although the editors requested that Sokal cut parts of the article, including nineteen pages of citations, in the end they accepted it unchanged for inclusion in their "Science Wars" issue, which purposed to examine the resistance of scientists to social constructivism. Strong social constructivism holds, after Michel Foucault, that "nature" itself is a social construction fraught with politics. The "Science Wars" issue framed itself as social constructivists' response to Paul Gross and Norman Levitt's 1994 book *Higher Superstition*, which criticized the sloppy appropriation of scientific concepts and terminology by cultural studies scholars.

On the same day that the "Science Wars" issue of *Social Text* came out, Alan Sokal published a companion article in *Lingua Franca* explaining that his *Social Text* article was a hoax. Sokal described it as an experiment: "Would a leading North American journal of cultural studies—whose editorial collective includes such luminaries as Fredric Jameson and Andrew Ross—publish an article liberally salted with nonsense if (a) it sounded good and (b) it flattered the editors' ideological preconceptions?"

The hoax became an overnight industry. The editors of *Social Text* retorted that they had not really been fooled after all, that they had read Sokal's article as an interesting "document" of resistance by scientists to cultural studies. That statement prompted a barrage of criticism. Scholars and lay people wrote into news media and academic journals alike voicing both support and disdain of Sokal's project, both criticism and support for cultural studies. The argument quickly exploded to the Internet and the mass media and became the topic of university forums and journal issues; similarly, the issues raised by Sokal's original hoax were lost in a free-for-all about the postmodern Left, the quality of college education, the role of women and non-Western others in the science wars, and the very issue of truth—if there were any such thing in the first place, and if there were, was science its only trustworthy oracle?

Reaction to the hoax has received more scholarly scrutiny than the hoax itself. The editors of *Lingua Franca* have published an entire volume of reactions to the Sokal hoax, and the reactions to those reactions, both in the domestic and foreign media. The issues of the science wars in America, the "two cultures" controversy, and the social constructivist approach to science studies have all been commented upon by eminently qualified scholars in both the sciences and arts. However, until Marie Secor and I published an article in *Written Communication* analyzing Sokal's rhetorical exploitation of the rhetoric of cultural studies, the rhetorical power of the hoax itself remained unaccounted for. This conclusion contributes three unique keys to understanding the Sokal hoax: first, I provide a portrait of the genre of cultural studies articles, much as I did for the nineteenth-century popular science article, and I show how Sokal explicitly exploited reader expectations about it; second, I describe why the hoax's rhetorical effect differed from, say, a critical article on the same topic or an exposé-style book like Gross and Levitt's; third, an analysis of the language used in reactions to Sokal's hoax illuminates the specific ways in which scientific media hoaxing has changed since the nineteenth century (and how it has remained the same).

Exploiting the Conventions of the Cultural Studies Article

Since hoaxes exploit readers' assumptions about genre, the first question to answer in an analysis of the effectiveness of Sokal's style with his audience is the following: what might readers of *Social Text* have expected from the typical cultural studies article in terms of its critical methods and ideologies, argumentative topoi, style, and citation practices? While not formally trained as a rhetorician, Sokal himself almost certainly conducted an intuitive rhetorical analysis of *Social Text*'s "style" before he submitted his hoax to the journal, as many academics do when they are considering submitting an article to a certain journal. I reenacted this work that Sokal likely did to acquaint himself with the conventions of cultural studies articles by randomly sampling eight full-length articles, one from each issue of *Social Text* published during the period when Sokal was constructing his hoax (1994–1996).[2] The eight articles were surveyed for recurring features under the following headings: critical/ideological stance, topoi, style, and citation. Even in this limited sample the following recurring themes or conventions evinced themselves strongly:

> **Critical/ideological stance**: The articles in the sample paid homage to the following schools, ranked from strongest to weakest influence: socialism/Marxism (following Stanley Aronowitz and Frederic Jame-

son particularly), French postmodernism and feminism, postcolonial theory, and science studies (dominantly the Edinburgh, French, and American feminist strains).

Topoi: The articles were structured around two major topoi that each had a Marxist or deconstructionist mode (sometimes both).

- "critiquing" or "interrogating" the use of a certain term, generally showing how it contains its opposite or how it participated in its own transcendence in a Marxist dialectic
- evaluating a new policy or cultural trend according to socialist/Marxist criteria or according to Foucaultian criteria of corroboration/resistance. In the latter mode the policy is usually found to buy into the current dominant power structure in some ways and resist it in others.[3]

Style: The style of the articles revealed patterns in syntax, wordplay, qualification, and jargon:

- complex syntax featuring the heavy nominalization typical of continental philosophy
- a marked emphasis on French postmodern word-play, especially hyphenation, slashing, and bracketing of alternative meanings
- constant hedging, deconstruction, and qualification of claims
- a core lexicon of jargon from the following fields: Foucaultian power dynamics (e.g., *hierarchy, dominant/domination, resistance, legitimate*); Marxist dialectic (e.g., *hegemony/hegemonic, internationalism, transcendent, global capitalism, emancipatory, solidarity, dialectic, totalizing*); French postmodern critical moves (e.g., *critique, homogeneous, heterogeneous, interrogate, discursive, signifier, bricolage, hyphenated, problematic, uncertainty*); postcolonial criticism (e.g., *globalization/global, localization/local, diaspora, Eurocentric, tribal*); science studies (e.g. *hybridity, objective, universalizing*); and Habermasian publics theory (*private, public, civil society*).

Citation: The authors of the articles in the sample repeatedly cited the following scholars (an asterisk denotes an editor or advisor of *Social Text* itself): Stanley Aronowitz*, Louis Althusser, Zygmunt Bauman, Jacques Derrida, Michel Foucault, Donna Haraway, Frederic Jameson, Andrew Ross*, Edward Said, Michael Shapiro, Ella Shohat*, and Gayatri Chandravorty Spivak. Seven of the eight articles in the sample acknowledged and/or cited a member of the *Social Text* editorial collective or advisory board (Aronowitz was cited in four of the eight articles). Three of the eight articles were written by *Social Text*'s editors or advisors.

With this profile of the typical *Social Text* article in view, we turn to the next question about Sokal's hoax: how well did Sokal do in making his article fit the *Social Text* profile and therefore in creating a "ringer" that would sucker the editors into reading it as a legitimate contribution? An examination of the rhetoric of Sokal's piece shows that he paid painstaking attention to each of the types of patterns discovered in the *Social Text* sample.

Critical/ideological stance:

- In his hoax Sokal cast himself as a leftist (which he is in real life) by claiming to embark on the project of creating a "liberatory" science that could resist and eventually supplant capitalist science.[4] This stance clearly fits the profile of the typical *Social Text* article.
- Further, Sokal forecasts a "truly progressive science" that will serve the "radical democratization of all aspects of social, economic, political, and cultural life."[5] Radical democracy is Stanley Aronowitz's principal political program.[6] Thus, Sokal made specific efforts in choosing his rhetorical ideology to flatter *Social Text*'s most important advisor.

Topoi: Sokal frames his article as an evaluation of the principles of the new science of quantum gravity according to radical democratic ideals for a progressive science.[7] This topos of evaluation is the most frequent high-level organizer in the sample of *Social Text*'s articles.

Style: Sokal paid close attention to the low-level rhetoric, syntax, and lexicon of *Social Text* articles in creating his hoax.

Syntax: Sokal used long, complex sentences with heavy nominalization, highly typical in *Social Text* pieces, as in this example from his opening paragraph:

> Rather, [natural scientists] cling to the dogma imposed by the long post-Enlightenment hegemony over the Western intellectual outlook, which can be summarized briefly as follows: that there exists an external world, whose properties are independent of any individual human being and indeed of humanity as a whole: that these properties are encoded in "eternal" physical laws; and that human beings can obtain reliable, albeit imperfect and tentative, knowledge of these laws by hewing to the "objective" procedures and epistemological strictures prescribed by the (so-called) scientific method.[8]

Qualification: As evinced by the above excerpt, Sokal frequently employed hedges and qualifications typical of cultural studies rhetoric. "(So-called) scientific method" is one example. A more complex qualification comes from his introduction, "It should be emphasized that this essay is of necessity tentative and preliminary; I do not pretend to answer all the questions that I raise."[9]

Wordplay: Sokal engages in wordplay that gives a nod to the French deconstructionist pedigree of *Social Text*. Some examples of this type of critical wordplay that reveals contradictions and reversals within seemingly unitary meanings include Sokal's repeated use of the term *(w)holism* to describe quantum gravity, as well as the following sentence: "Suffice it to say that anyone who has seriously studied the equations of quantum mechanics will assent to Heisenberg's measured (pardon the pun) summary of his celebrated uncertainty principle."[10]

Jargon: Sokal whips out all of the cultural studies jargon, including but not limited to the following major buzzwords: *hegemony*, *domination*, *emancipatory*, *dialecticism*, *deconstruct*, *transcend*, *ideology*, *capitalistic*, *problematize*, and *transgressing*. Significantly, he also used scientific jargon, like "Planck-scale" and "open strings," that was probably unfamiliar to cultural studies scholars—just as nineteenth-century scientific hoaxers such as Poe and Locke laid an impenetrable veneer of astronomical jargon over their hoaxes in order to lend them an air of scientific authority.

Citation: Sokal clearly did his homework on who was involved in and fashionable with the *Social Text* editorial collective.

- Among a ludicrously weighty nineteen pages of notes and references (which *Social Text* editors unsuccessfully requested that he cut down),[11] Sokal cites Marxist and socialist critics Althusser, Aronowitz, and Jameson; French postmodernists Deleuze and Guattari, Derrida, Irigaray, Serres, Lacan, and Lyotard; science studies critics Haraway, Harding, and Latour; and, in addition to Aronowitz, *Social Text* editor Andrew Ross.
- Sokal went beyond simply structuring a skeleton of typical cultural studies references; he fleshed it out in the text of the hoax by making bold, sweeping claims capped off with parenthetical citations—a convention of *Social Text* articles—thus forcing his readers either to view the notes or look up the cited studies if they wished to see the actual data supporting the claims. For example, in his discussion of differential topology and homology, Sokal structures his support as follows: "Furthermore, as Lacan suspected, there is an intimate connection between the external structure of the physical world and its inner psychological representation qua knot theory: this hypothesis has recently been

> confirmed by Witten's derivation of knot invariants (in particular the Jones polynomial [Jones 1985]) from three-dimensional Chern-Simons quantum field theory (Witten 1989)."[12] Sokal does not explicate knot theory or any other technical terminology, nor does he show how they support his argument in an endnote. Eliding the data and interpretation from this support structure is convenient shorthand but blocks nonexpert readers' ability to evaluate his claims until they have had the opportunity to look up Jones's and Witten's studies.

The excellent fit of Sokal's rhetorical strategies to the *Social Text* paradigm suggests that he made explicit efforts to satisfy editorial expectations. That the hoax was accepted for publication is of course not necessarily proof that these strategies were the sufficient cause of its acceptance.[13] However, the hoax's rhetorical mimicry of the typical *Social Text* article certainly encouraged the editors' reception of it as a standard contribution to their journal. In particular, Sokal exploited the cultural studies convention of "shorthanding" citations—providing the reader only a bald claim of support and a citation while eliding the data and interpretation that connects them—to work in dozens of nonsensical claims that the editors never took the time to verify. He hoisted them on their own conventions.

Sokal's Hoax Constructs Him as a Notorious Expert

The scientific jargon Sokal layered onto his hoax constituted his most interesting trick and a key point for our discussion of the effects of his hoax: this trick was his version of De Quille's "Emperor's New Clothes" appeal. He dressed his article in scientific terminology that the *Social Text* editors could not admit ignorance of without appearing to be illiterate in a discipline that several of them regularly critiqued.

Fascinatingly, many of the media reactions to the hoax (and Sokal's own revelation of it) referenced exactly the folk tale in question, but from the perspective of the bystanders who were watching the emperor parade by them in his birthday suit. Many critics felt cultural studies had finally been "exposed" by Sokal's hoax for the intellectual charade and waste of taxpayer money that they felt it had always been. However, there is another angle to the "Emperor's New Clothes" story, the angle De Quille also exploited in his hoaxing, and that is the tailor's point of view. Sokal, as the canny tailor, casts the editors of *Social Text* in the emperor's role. Unwilling to admit that they really did not know enough science to judge the merits of Sokal's quantum mechanical argument, they published his article on the grounds that it was better to risk fallout than to lose the chance to have a "real" scientist supporting cultural studies.[14] The editors' own statements corroborate this reading to some extent. In their defense of their acceptance of Sokal's article, they claimed that because they "try to keep abreast of cultural studies," they viewed Sokal's contribution as "unusual," coming from a natural scientist, and therefore "worth encouraging."[15] This focus on maintaining face within a particular community sometimes detracts attention from facts and details, and that is exactly the same psychological response De Quille counted on in his readers when he employed his "anyone who knows anything" strategy for securing belief in his hoaxes.

Another familiar feature of Sokal's hoax is the public notoriety it created for him as a countercultural leftist fighting, in the view of some, the excesses of the culture-studies Left and, in the view of others, the entire bloated edifice of the academic humanities. There is at least some evidence that this notoriety was a partial goal in constructing his hoax. First, there is the evidence of the hoax itself. In the *Lingua Franca* piece, Sokal defended attacks on his ethics by saying he chose "satire" as a weapon because it was "an attack that could not be brushed off."[16] Sokal's choice of "satire" rather than "hoax" was probably deliberate and is highly significant; its meaning(s) will be discussed in a moment. But, the very reason a hoax cannot be "brushed off" is because it constructs a Barnumesque public spectacle beyond the control of its victims: the spectacle of a victorious, notorious expert versus losing dupes. If Sokal had not intended to construct this ethos for himself, he would have worked through conventional channels to voice his criticisms of the *Social Text* crowd directly in an academic essay.

Further evidence of Sokal's desire for notoriety (or at least his willingness to accept it, as Poe accepted the role of notorious expert in responding to queries about M. Valdemar) comes from Sokal's active role in perpetuating the public life of the hoax. He wrote hundreds of emails in its wake, attended public forums on it, wrote response pieces for national and international journals. All in all this campaign cost him a three-year black hole in the "Publications" section of his physics CV. In 1997, just as it appeared the hubbub was about to die down, he published a book with European physicist Jean Bricmont titled *Les Impostures Intellectuelles*, in which he extended arguments about the inadequacy of French postmodernist critiques of science that he had begun making during the reaction to his hoax. Finally, there is the evidence of the evolving scope of Sokal's arguments over time. He began, in *Lingua Franca*, insisting he only had a small "target" in mind with the hoax, the editors of *Social Text*. But as reader reaction expanded his role and his position, crediting him with nothing short of an attack on culture studies as a whole for its role in diminishing the public reputation of American science, Sokal stepped into these larger and larger shoes with such alacrity that it seemed that this was where he intended to go with his hoax in the first place. In a public lecture given at a New York University forum just a few months after the hoax was published, Sokal unveiled this expanded vision:

> *Social Text* is not my enemy, nor is it my main intellectual target. . . . Rather, my goal is to defend what one might call a scientific worldview—defined broadly as a respect for evidence and logic, and for the incessant confrontation of theories with the real world; in short, for reasoned argument over wishful thinking, superstition and demagoguery. And my motives for trying to defend these old-fashioned ideas are basically political. I'm worried about trends in the American Left—particularly here in academia—that at a minimum divert us from the task of formulating a progressive social critique, by leading smart and committed people into trendy but ultimately empty intellectual fashions, and that can in fact undermine the prospects for such a critique, by promoting subjectivist and relativist philosophies that in my view are inconsistent with producing a realistic analysis of society that we and our fellow citizens will find compelling.[17]

It seems clear from these comments that Sokal's ultimate goal was a broad intellectual stance, and the hoax genre afforded him the ideal public stage on which to take it.

As counterevidence for these claims about Sokal's intentionality with respect to the construction of his public ethos, one might observe that Sokal initially tried to distance himself from the notoriety the hoax constructed for him and the aspersions it cast on him ethically. As mentioned above, in his *Lingua Franca* revelation, he eschews the term *hoax* altogether, preferring the more literary *parody* and *satire*, friendlier critical genres that place author and audience on the same side against a cultural enemy, as noted above. In later commentary, however, including the New York University forum speech excerpted above, he began owning his original article as a hoax, and this shift in terminology corresponds with his acceptance of a larger public role as a spokesperson for the natural sciences against the incursion of cultural studies. It remains to be seen if the ethics of the hoax will infect Sokal's reputation as an academic, as in the case of Poe and Twain, or if, as in De Quille's case, Sokal's physics is a separate enough endeavor from his social commentary that his scientific reputation will remain unsullied among his peers. Certainly, humanists such as Stanley Fish have argued that Sokal's scientific ethics are suspect because of his media ethics.

But Jeanne Fahnestock and Marie Secor promptly countered Fish by arguing that Sokal's projects are ethically as well as topically separate. Further, since Sokal attacked members of a rival field, his reputation among particle physicists and other scientists may remain unsullied or may perhaps be strengthened because these colleagues are "on his side."[18] The test of time, as always, will determine Sokal's historical reputation as a scientist in light of his stint as a hoaxer.

The Hoax as a Computer Virus

Setting these familiar features of Sokal's hoax aside—its construction of notoriety, its dependence on jargon and citation to create verisimilitude—crucial rhetorical innovations distinguish it from the nineteenth-century hoaxes. First, Sokal is a scientist creating a literary hoax for cultural studies scholars, which is an inversion of the nineteenth-century scientific media hoax dynamic. Now, it is our scientists, not our artists, who feel their social prestige is at stake. Hoaxes remain a subversive strategy for taking the stuffing out of newly prominent people or institutions. The commentary in the Sokal affair figures literary scholars as the pretenders to the throne whom scientists are desperately trying to disenfranchise. This reactionary rhetoric is backed up by economic metaphors in the reactions to the hoax. The issue of the public funding of science keeps coming up: some critics suggest Sokal and his fellow physicists are sniffing around for a new enemy after the Cold War and the collapse of the defense industry;[19] others suggest that Sokal is bitter because cultural studies and science studies are starting to dampen the American enthusiasm for science—an enthusiasm based on science's spectacular performances for America in the World Wars and the Cold War.[20] Both of these views paint Sokal and his colleagues as threatened by the incursion of culture studies—the exact inverse of the picture that Poe's and Twain's hoaxes painted of artists being threatened by scientists.

Another significant difference in Sokal's hoax lies in the audiences it invoked. It was not, like our nineteenth-century hoaxes, a story in a newspaper about a "fake" event—cultural studies scholars finding evidence that things disappeared when they stopped thinking about them, for instance, or proving that gravity was propaganda. Instead, Sokal's hoax began by targeting a very small professional readership—the editors of *Social Text*. They were the audience who had to make the crucial decision about the truth or falsehood of the text in front of them. However, because it is the job of those editors to publish the results of those value judgments, the hoax had a second life, so to speak, in which both it and Sokal's revelation of it appeared simultaneously to a different and much wider readership. For this second readership, Sokal's article actually read as the "satire" or "parody" that he initially named it, since these readers were in on the joke rather than on the business end of it. This crucial distinction between readerships helps explain why there is so much terminological confusion in the reaction to Sokal's hoax. The editors of *Lingua Franca* sum up the mayhem cleverly: "In the headlines alone, Sokal's article has been called a hoax, a joke, a sting, an affair, a *paródia*, a prank, *uno sfregio*, a spoof, a con, *un canular*, a fraud (delicious and malicious) a ruckus, *la farce parfaite*, a Pomolotov Cocktail, a *brincadeira*, a mystification pédagogique, double-speak, *un'atroce beffa*, nonsense, gibberish, rubbish, and hokum."[21]

Many of these variants, of course, refer to differing aspects of the Sokal affair and amount to scholarly word play. However, there is one rhetorically significant confusion of terminology, even in Sokal's own discussion of his hoax—the confusion between hoax and parody. The article's nested audiences (first, the "hoaxed" editors and, beyond that, the general public enjoying the "parody") legitimates both terms, even though no one involved in the controversy clarified this point. Many commentators do demonstrate an unconscious awareness of the difference: criticisms of Sokal's agency in the matter tend to favor the word *hoax*, especially as a verb; those commentators who discuss the article's literary aspects or who efface Sokal's agency tend to prefer *parody*. Sokal himself uses both terms eventually, but in the initial revelation of the hoax in *Lingua Franca* in 1996, Sokal carefully avoids using *hoax*. He uses *parody*, *satire*, *spoof*, and *experiment*, and he accompanies these words with corroborating words like *silliness* and *nonsense*. As we discussed with respect to author intentionality, *hoax* carries with it a connotation of villainous agency. In his revelatory article, Sokal is clearly trying to impose a level of artistic and aesthetic distance between himself and his deliberate attack on the reputation of the editors of *Social Text*.

A final difference between the Sokal hoax and the nineteenth-century hoaxes is both a difference and a similarity. The Sokal hoax is still patterned after a salient technology in its reading culture—but that is now the computer virus, rather than the gears-and-girders-type machine. Kaufer and Carley argue that the virus is now the major metaphor for communication among researchers in the field.[22] The computer virus, a hybrid of mechanical and organic villainy, is therefore the perfect double for the rhetorical mechanics of the twenty-first-century hoax. Far from being a clever argumentative convenience, this metaphor actually helps explain key aspects of Sokal's hoax: his choice of the hoax as a mode of attack, his hoax's dependence on the Internet, and his choice of *Social Text* as his target.

The connection between the computer virus and the Sokal hoax is apparent if we consider the language of reaction to the hoax, the media of transmission and reaction, and the dominant mode for hoaxing in the 1990s.

Sokal himself uses the words *deception*, *weapon*, and *attack* in his *Lingua Franca* piece to describe his project. The reactions to the hoax give it all sorts of labels, but two pervasive metaphors (aside from the emperor's new clothes) pick up on Sokal's own assessment: *terrorism* and *mechanics/technology*. The hoax is referred to as "intellectual terrorism," and Sokal is compared to Ted Kaczynski; these references are accompanied by references to bombs and explosions, including words such as *defuse*, *detect*, *breach*, *attack*, *burn*, and *fireworks*. The other persistent analogy is to mechanism, particularly to a trap. The mechanical language includes words such as *picks up steam*, *contrived*, and *fabricated*. The hoax is elsewhere figured as a *trap* that was *camouflaged* for *prey* that *took the bait* or *lure*.

The common denominator of terrorisms, traps, and technology is not immediately apparent unless we consider a major player in the Sokal affair that has been almost entirely overlooked in the voluminous commentary—the Internet. Peter Osborne was the only major commentator who even mentioned that the internet "played a significant role in framing and sustaining the affair,"[23] although he did not go into much detail about this role. He noted that much of the discussion after the hoax's publication took place in emails and Internet forums between the principals and observing critics, which is significant. However, he did not mention that the crucial negotiation of the placement of Sokal's hoax in *Social Text* took place between the editors and Sokal via the Internet. Nor did he mention that to this day the hoax is sustained by a virtual reconstruction of all of the original texts and texts of reaction on Sokal's personal web page, where it shares space with propaganda about *Impostures Intellectuelles* and commentary of other natural scientists criticizing cultural studies.

I argue that the strong Internet presence constructed by Sokal for his hoax is not coincidental. Sokal is exploiting the primary medium of twenty-first-century hoaxing. From "how to make your own atomic bomb" websites to fake web diaries of teenagers with STDs to "urban legend" and scam emails, the Internet has picked up where nineteenth-century print media left off. Objectivity standards in print and television media have made them inhospitable (while not impervious[24]) to hoaxing; meanwhile, the development of the Internet offers an attractive alternative medium for hoaxers. First, its distributed network resists monitoring by any central authority and therefore frustrates would-be censors and referees. Print publishing houses, doing business in the physical rather than the virtual marketplace, are subjected via their physicality to a higher degree of centralized control; there is somewhere you can "go" to stop a story or interrogate a writer. On the Internet anyone with access to a server can publish and disseminate any message she wishes. Further, this lack of centralized authority makes the construction of Internet ethos difficult and interesting. How does an anonyomous email or webpage accrue authority? In the absence of personal information or references, resources such as graphic presentation, timeliness of topic, celebrity, language, and ubiquity play a powerful role in establishing the credibility of information received via the Internet. Interestingly, if we look at our filters from the nineteenth century, a similar result can be obtained by stripping off most of the top tier of expectations—those associated with the reputation of the author and the medium. In the absence of those powerful deciding factors, corroboration, plausibility and novelty, visual impact, format, and language determined interpretation for Poe's, Twain's, and De Quille's readers—and for present-day Internet readers. The media revolution represented by the Internet can be characterized as a diminishment of the resources readers have at their disposal to guide judgments about truth and falsity.

In spite of this anarchic picture, the Internet is hardly a den of thieves. Trustworthy institutions and foundations also make use of the Internet to stay in convenient contact with their clients. Therefore, enough useful and credible information comes via the Internet that it presents a very similar environment to the penny daily of the 1830s. A carrier of both fact and fiction, it is the perfect milieu for hoaxes.

The most common type of Internet hoax is the email hoax. It is very quick and anonymous (with the use of fake names and email accounts), and has the added bonus of accumulating ethos for the hoax through the headers of concerned friends who forward the warnings about gang initiations, fast food contaminations, and computer viruses. The computer virus scam email is the most interesting of these Internet hoaxes for our purposes because it actually performs the fear of contracting the virus—"I got this email, so I could get the virus just as simply." Real computer viruses, of course, are easy to contract via the Internet; the Klezworm virus is a relatively recent example. It sends itself as a message with a vague lure of a subject header, such as "per your request," and when it is opened, it finds every confidential file in your computer and attaches it to emails that go out to everyone on your email list.

Sokal, likely unconsciously, adopted the signature of this salient technology to pattern his hoax. Like a computer virus, his hoax contained "lures" that tempted the editors of *Social Text* (called appropriately in some commentary the "host" journal of the hoax) to include it in the "Science Wars" issue. However, Sokal's simultaneous revelation in *Lingua Franca* made *Social Text* disseminate a far different message from the one they intended to communicate with that issue. The other articles in the issue were practically forgotten, and instead the attention of *Social Text*'s readers was focused on the gullibility of the editors and on Sokal's indictment of cultural scholars for their sloppy handling of scientific concepts. The sensation caused the "Science Wars" issue to sell out. In addition, the hoax disseminated over the Internet to reach universities and popular media, who then reprinted Sokal's basic message ("Cultural studies are ignorant and politically dangerous; science is still the only objective channel to truth and reality") to a wide lay audience.[25] The essential functions of the computer virus—luring a host machine into reproducing and disseminating the virus to other machines—are all performed in Sokal's hoax.

A computer virus damages the host and uses the host to spread its code, its message, across the Internet, and both this damage and this dissemination seem to be Sokal's intention. Osborne comments, "Sokal has used the media skillfully, both to register his hoax and to generalize its point into a full-scale attack on 'cultural studies of science' and 'postmodern cultural studies' (which he tends to treat as equivalents)."[26] Just as computer virus hoax emails perform the danger of the "viruses" they warn users about, Sokal's virus hoax performed the vulnerability of cultural studies to attack by any discipline whose discourse they used without true understanding. Sokal's specific goals dictated his choice of the hoax over other methods of public criticism. He explained his motivations in his revelation:

> In the end, I resorted to parody for a simple pragmatic reason. The targets of my critique have by now become a self-perpetuating academic subculture that typically ignores (or disdains) reasoned criticism from the outside. In such a situation, a more direct demonstration of the subculture's intellectual standards was required. But how can one show that the emperor has no clothes? Satire is by far the best weapon; and the blow that can't be brushed off is the one that's self-inflicted.[27]

Sokal's "Emperor's New Clothes" reference is doubly significant to a final understanding of his hoax as a twenty-first-century scion of the tradition of hoaxing scrutinized in this book. First, the reference implies that print media, when compared to the Internet that sustained much of "*l'affaire Sokal*," is still the "emperor" of media. The academic journal is the prestige form against which the cyber-manifestations of Sokal's hoax—including his email negotiations with the editors, his online debates, and his personal webpage that documents and reifies the hoax—constitute a guerilla-style assault. A second but equal significance of Sokal's "Emperor's New Clothes" allusion is that it references the primary social function of hoaxing, which Sokal exploited even if unconsciously: the hoax's ability to call into question the construction of reality. The hoax, as we have discussed, operates at the stasis of existence; it plays with what holds true in the world views of its readers. An evaluation can be drawn from it, and most of the discussion of the hoax has been absorbed in evaluation, as Sokal intended. But the basic function of the hoax is to call into question assumptions about the real world. This is what made it the ideal mode of attack for his purposes. Sokal positioned himself as defending objective scientific reality from the attacks of the "emperors" who claimed there was no objective reality.[28] What better way to make his point, then, than to use a hoax to show that the world view of the cultural studies "emperors" amounted to a lot of thin air and that scientists, in the final accounting, are the tailors—with their eyes wide open and their purses full of the emperors' money.

Appendix A: How to Read Tables in Optimality Theory (OT)

Table A1.1 is a representation in OT of how English speakers unconsciously select the optimal syllabification of the word *onset* (/ansɛt/ in International Phonetic Alphabet notation).

What the Parts of the Table Mean

- The leftmost column lists the most probable syllabifications of the English word *onset.* A hyphen indicates the syllable break in the word. The bracketed <a> in the third candidate represents a deleted vowel (a fairly common phonological feature in colloquial English: think of "N-n" for "No, no").
- The top row lists all phonological constraints that apply to syllabification left-to-right from strongest (inviolable) to weakest (often violated in practice). The ranking was predetermined from analyzing many phonological data sets in English. The

Table A.1 Syllabification of /ansɛt/ in OT

	FAITH	ONS	NOCODA
✓an-sɛt		*	**
Ans-ɛt		*	*!**
<a>nsɛt	*!		*

constraint FAITH requires that all parts of a word should be pronounced; ONS says syllables should start with consonants; and NOCODA says syllables should not end in consonants. The solid vertical line between FAITH and ONS divides the levels of ranking and tells us that FAITH crucially dominates both ONS and NOCODA. ONS and NOCODA are unranked with respect to each other because they never operate on the same part of the syllable and therefore never compete with each other; the dotted line signifies this lack of competition. The ranking of the constraints in this tableau could also be notated in a linear form as FAITH>> {ONS, NOCODA}, where ">>" signifies domination, and bracketing with commas signifies equality of rank and therefore lack of competition.

- The asterisks in the matrix of the table represent violations of particular constraints. The violations add up like penalty points against a candidate, with a violation of a stronger (leftward) constraint counting more than a violation of a weaker one. An "!" follows and indicates the fatal violation, the one that knocks the candidate out of the running for optimal form (violations are usually counted up from right to left, weakest to strongest). A candidate can earn a fatal violation either by violating a higher-ranked constraint than the other candidates do, or by accumulating more total violations than other candidates at the same level of ranking.
- The check mark in the candidate column indicates the optimal candidate phonological form, the one that "wins" by accumulating the fewest violations of higher-ranked constraints. This is the form speakers actually use when they pronounce the word *onset.*

The Results of the Syllabification of /ansɛt/

In the example in table A.1, "an-sɛt" is the optimal form. While it has more total violations than "<a>nsɛt," it satisfies FAITH, the highest-ranked constraint. The runner-up, "<a>nsɛt," does not. The third form, "ans-ɛt," gets knocked out of the running even earlier because it accrues more NOCODA violations than either of the other two forms due to a consonant cluster "ns" at the end of the first syllable.

Optimality Theory Applied to a Decision about a Hoax's Truth-Value

Table A.2 represents a decision about the truth-value of Richard Adams Locke's Moon Hoax by a reader whom Poe projects to value spectacle over strict technical consistency and accuracy (see chapter 3 for a history of this hoax).

Table A.2 A Reader's Decision to Believe the Moon Hoax Based on Spectacle, Not Science

	Novelty	Sensation	Popsci.	Plausibility	Internal Coh.
✓ TRUE				********	*
FALSE	*	*	*!		

The bold vertical line locates the conflict—between first impressions and factuality, essentially. The dotted lines denote a lack of evidence for competition in this particular decision about the truth of the hoaxes; that is, novelty, sensation and popsci. (verisimilitude) "work together" rather than compete with each other as the reader reads.

In the graph the nine factual errors in Locke's story are counted with nine asterisks (representing eight violations of plausibility and one of internal coherence. In spite of all of this faulty evidence, Poe represents this reader as still believing Locke's story to be "true" because to consider it "false" would force the admission that something novel, sensational, and "verisimilar" is not true. The exclamation point on the chart indicates that all violations at this highest level of expectation are unacceptable (the convention is to mark unacceptability on the very first violation that renders the candidate interpretation unacceptable, counting violations right to left). Thus, the candidate with more total violations actually wins in this case because of the very low value assigned to scientific accuracy by the reader.

Appendix B: Reader Responses to the Moon Hoax

NEWS FROM THE MOON.—The **news** from this planet, received through the Sun, is truly **astonishing**, and I look with as **much anxiety** for the finale, as the public do for the receipt of the 2d volume of Japhet in search of his father. The man Bat, alluded to, will evidently be brought forward to prove the Mosaic account of the Creation, and I expect ere long **Sir John Herschel** will discover a long tail, but whether he will have 'breeches of blue' is a matter of doubt. Who Sir Johu [sic] Herschel is, I do not know; it <u>cannot</u> be Sir John F. W. Herschel, who is now at the Cape of Good Hope, as he would not be so <u>ignorant</u> as to suppose, **that he could by any concentric circles, trace the shadow of any heavenly body on the earth's surface, unless the observatory moved on a plane parallel to the equator**. Had he described an equatorial plane erected for the lat. of the Cape of Good Hope, he might have been able to keep his observatory within the focal distance, provided his object glass had also an equatorial movement.

I would recommend the articles be re-published, and bound up with Gulliver's Travels, and the Voyages of Peter Wilkins, and that the editor of the Sun should secure a copy-right.—A Terrestrial.

—N.Y. *American*, 28 August 1835

No article, we believe, has appeared for years, that **will command so general a perusal and publication. Sir John** has **added a stock of knowledge** to the present age that will immortalize his name, and place it high on the page of science.

—N.Y. *Daily Advertiser*, 28 August 1835

DISCOVERIES IN THE MOON.—We commence to-day the publication of an interesting article which is stated to have been copied from the Edinburgh Journal of Science, and which made its first appearance here in a cotemperary [sic] journal of this city. It appears to carry **intrinsic evidence** of being an authentic document.

—*Mercantile Advertiser*, quoted in Ormond Seavey's introduction to *The Moon Hoax: Or, A Discovery that the Moon has a Vast Population of Human Beings*

STUPENDOUS DISCOVERY IN ASTRONOMY.—We have read with **unspeakable emotions of pleasure and astonishment**, an article from the last *Edinburgh Scientific Journal*, containing an **account** of the recent **discoveries** of Sir John Herschel at the Cape of Good Hope.

—*Albany Daily Advertiser*, excerpted from Seavey

It is quite proper that the *Sun* should be the means of shedding so much light on the *Moon*. This singular narrative of lunar **discoveries** purports to be extracted from a supplement to the Edinburgh Journal of Science, with which the editor of the *Sun* says he has been 'politely furnished by a medical gentleman immediately from Scotland.' We publish the article as we find it, and do not know that it is necessary that we should accompany it with any comments to shake the faith which credulous readers maybe disposed to place in its authenticity. The story is certainly, as the old newspaper phrase goes, "very important, if true." And if not true, the reader will still be obliged to confess that it is very **ingenious**. The **account** of the *Vespertilio homo*, or *Man-bat*, which will be found in the latter portion of the extract, is not the least **remarkable** part of the narrative. But that there should be winged people in the Moon does not strike us as more wonderful than the existence of such a race of beings on earth; and that there does or did exist such a race rests on the evidence of that most veracious of voyagers and circumstantial of chroniclers, Peter Wilkins, whose celebrated work not only gives an account of the general appearance and habits of a most interesting tribe of flying Indians, but also of all those more delicate and engaging traits which the author was enabled to **discover** by reason of the conjugal relations which he entered into with one of the females of the winged tribe. The **particulars** related, also, of the obelisks of amethysts and caverns of precious stones, are not more strange than the details given by Sinbad the Sailor of the mines of diamonds, in the dazzling chambers of which he was thrown, after suffering a long time in a most dreadful storm at sea, and undergoing various adventures on land, hardly paralleled by those related of Aeneas by Virgil. We have no wish to depreciate the lunar **discoveries of Mr. Herschel**, however, by referring to these **mun-**

dane circumstances, and commend the article to our readers as one from which they will derive much **entertainment** whether they wholly believe it or not.

—*N.Y. Evening Post*, 28 August 1835

The town has been agape two or three days at the very ingenious astronomical hoax, prepared and written for the Sun newspaper, by Mr. Locke, formerly the police reporter for the Courier and Enquirer . . . [who] has struck out a piece of invention of superlative drollery, which has actually deceived several of our men of science (have mercy on us.) He has, however, dressed it up, in rather too much finery—and made **several mistakes in his philosophical keeping**. His descriptions of the shadows in the moon are **incorrect on mathematical principles**, and we also **doubt whether his optical principles correspond with fact**. He has also made a **blunder in the name of Herschell**, dubbing him an LL.D. which is not the fact—**also in calling the supplement the 'Edinburgh Journal of Science' no work of that kind being now published**.

—N.Y. *Herald*, 31 August 1835

In this age of **wonderful discoveries**, none has struck the mind with such **astonishment**, as those that have resulted from the lunar observations of **Sir John Herschell**, recently published in the Supplement to the "**Edinburgh Journal of Science.**" Indeed, so **surprising** are these discoveries, that a few persons in this city—and we believe a very few—actually doubt—or affect to doubt—the whole story, and declare that they should as soon credit the story of that renowned master of fiction, Capt. Lemuel Gulliver himself.

The account, we confess, is **marvellous**, but not therefore necessarily **false**. Indeed, so great are the discoveries of modern science, that it is evidently rather a proof of wisdom, good sense, and philosophy, to believe *every* thing, however extraordinary and **apparently improbable**, than to pretend to doubt of *any* thing.

—N.Y. *Transcript*, 29 August 1835

THE GREAT MOON STORY.—At the request of many friends, and partly also for its ingenuity, we have copied on our first page the **marvellous** account which has been made the occasion of so much discourse lately of certain **stupendous discoveries** in the moon, stated to have been made by **Sir John Herschel** from his observatory at the Cape of Good Hope. It is **well done**, and makes a **pleasant piece of reading** enough, especially for such as have a sufficient stock of available credulity; but we can hardly understand how any man of common sense should read it without at once perceiving the **deception**. Without referring to the **monstrosities of the story itself**, can any one suppose for a moment that such preparations as are described, should have been made **without a word of notice in the English papers**? Preparations going on for years—an object glass of twenty-four feet in diameter—a donation of ten thousand pounds by the king—consultations with **Sir David Brewster**—and other extravagancies not less preposterous! **Wonderful** as is the tale, it is surpassed in **marvellousness** by the fact that it has found credence.

We are assured that the article was written in this country, and if so the **ingenious** perpetrator must be hugely delighted with its success; but we think we can trace in it marks of **transatlantic origin.**

—*N.Y. Commercial Advertiser*, 29 August 1835

The **writer** (Dr. Andrew Grant) displays the most extensive and accurate knowledge of astronomy, and the **description of Sir John's recently improved instruments**, the principle on which the inestimable improvements were founded, the **account** of the **wonderful discoveries** in the moon, &c., are all **probable and plausible**, and have an air of **intense verisimilitude**.

—N.Y. *Times*

GREAT ASTRONOMICAL **DISCOVERIES**!—By the late arrivals from England there has been received in this country a supplement to the Edinburgh Journal of Science containing intelligence of **the most astounding interest** from Prof. Herschel's observatory at the Cape of Good Hope. . . . The promulgation of these **discoveries creates a new era** in astronomy and science generally.

—*New Yorker*

Our enterprising neighbors of the *Sun*, we are pleased to learn, are likely to enjoy a rich reward from the late lunar **discoveries**. They deserve all they receive from the public—"they are worthy."

—*N.Y. Spirit of '76*, quoted in Seavey

After all, however, our doubts and **incredulity** may be a wrong to the **learned astronomer**, and the circumstances of this **wonderful discovery** may be correct. Let us do him justice, and allow him to **tell his story** in his own way.

—*N.Y Sunday News*, quoted in Seavey

The article is said to be an extract from a supplement to the *Edinburgh Journal of Science*. It sets forth difficulties encountered by Sir John, on obtaining his glass castings for his great telescope, with magnifying powers of 42,000. The **account**, excepting the magnifiying [sic] power, has been before published [i.e., in the Supplement to the *Edinburgh Journal of Science*.—Ed. *Sun*].

—*U.S. Gazette*, quoted in Seavey

It is not worth while for us to express an opinion as to the truth or falsity of the narrative, as our readers can, after an attentive perusal of the whole story, decide for themselves. Whether true or false, the article is **written with consummate ability**, and possesses **intense interest**.

—*Philadelphia Inquirer*, quoted in Seavey

In sober truth, if this account is true, it is most enormously **wonderful**, and if it is a **fable**, the **manner of its relation**, with all its **scientific details, names of persons employed**, and the beauty of its glowing descriptions, will give this ingenious history a place with "Gulliver's Travels" and "Robinson Crusoe."

—Philip Hone, former NYC mayor, quoted in Seavey

A **great talk** concerning some **discoveries** in the moon by Sir John Herschell; not only trees and animals but even men have been **discovered** there. It is all a hoax, although the **story is well put together**. The author of these **wonders** says that an enormous lens of 30 feet diameter was constructed. He thought that would be a big enough lie in all conscience, but he should have said **a lens of 100 feet diameter, as it is shown by writers on optics** that such a diameter would be required to ascertain if any inhabitants in the Moon. Why not make a good lie at once? But it is utterly impossible to construct a lens of half that diameter, and therefore we may despair of ever ascertaining whether the moon be inhabited.

—Michael Floy, Jr., quoted in Seavey

The singular **blunders** to which I have referred being properly understood, we shall have all the better reason for wonder at the prodigious success of the hoax. Not one person in ten discredited it, and (strangest point of all !) the doubters were chiefly those who doubted without being able to say why—the ignorant, those uninformed in astronomy, people who would not believe because the thing was so novel, so entirely "out of the usual way." A grave professor of mathematics in a Virginian college told me seriously that he had no doubt of the truth of the whole affair! The great effect wrought upon the public mind is referable, first, to the **novelty** of the idea; secondly, to the **fancy-exciting and reason-repressing** character of the alleged discoveries; thirdly, to the **consummate tact** with which the deception was brought forth; fourthly, to the exquisite **vraisemblance** of the narration.

The hoax was circulated to an immense extent, was translated into various languages—was even made the subject of (quizzical) discussion in astronomical societies; drew down upon itself the grave denunciation of Dick, and was, upon the whole, decidedly the greatest hit in the way of **sensation**—of merely popular sensation—ever made by any similar fiction either in America or in **Europe**.

—E. A. Poe, in his "Literati of New York" sketch of Richard Adams Locke, *Godey's Ladies' Book*, October 1846

[S]o thoroughly was the popular mind, even among the best educated and most reading classes, imbued with these fanciful anticipations of vast lunar **discoveries**, and of speedy telescopic **improvements** by which they were to be assuredly achieved, that, at the time Mr. Locke's "Moon Story" was written, scarcely any thing could have been **devised and announced upon the subject too extravagant for general credulity to receive.** To every observant advocate of legitimate science and authentic knowledge, this humiliating state of things must naturally have seemed to demand some pointed and practical reproof, to effect its exposure, correction, and reformation. Educated in the strictest school of mathematical and inductive science, zealously devoted to its studies and pursuits, and, at the same time, as we have already intimated, holding an influential connection with a **new** and vigorous branch of the public press, then rapidly acquiring an unprecedented breadth of popularity, Mr. Locke was precisely the writer to check so mischievous an epidemic of imaginary and spurious philosophy, by a well-timed satire, "out-heroding Herod" in its imaginative creations, supplying to satiety **the morbid appetite for scientific wonders then universally raging**, and, at the same time, maintaining a pretty firm grasp upon the credulity of nearly all but the thoroughly taught minds, by its **plausible display of scientific erudition**.

—William Griggs, in *The celebrated 'moon story,' : its origin and incidents; / with a memoir of the author, and an appendix, containing, I. An authentic description of the moon; II. A new theory of the lunar surface, in relation to that of the earth*, 1852

I happened to be going the round of several Massachusetts villages when the **marvellous** account of Sir John Herschel's **discoveries** in the moon was sent abroad. The **sensation** it excited was wonderful. As it professed to be a republication from the **Edinburgh Journal of Science**, it was some time before many persons, except professors of natural philosophy, thought of doubting its truth. . . . [M]y experience of the question with regard to the [moon story], "Do you not believe it?" was very extensive.

—Harriet Martineau, in *Retrospect of Western Travel*, 1838

The *Albion* from 29 August 1835 mentions as suspicious the lack of corroboration for the story and the improbability of getting an "exclusive-?? copy" of Grant's report, as well as the cost of the cost of the telescope being given in dollars rather than pounds.

The *Journal of Commerce* reprinted the story entitled "Hoax" on 28 August 1835.

We are necessarily compelled to omit the more **abstruse and mathematical parts** of the extracts however important they may be as a **demonstration** of those which we have marked for publication; but even the latter cannot fail to **excite** more **ardent curiosity** and afford more **sublime gratification** than could be created and supplied by any thing short of a direct revelation from heaven.

—New York *Sun*, 25 August 1835

Notes

Introduction

1. Samuel Langhorne Clemens, "Memoranda: A Couple of Sad Experiences," *Galaxy* 9 (1870): 858–61.

2. Fred Fedler, *Media Hoaxes* (Ames: Iowa State University Press, 1989), 40.

3. Hugh Kenner, *The Counterfeiters: An Historical Comedy*, reprint ed. (Baltimore: Johns Hopkins University Press, 1985; reprint, 1985), 30.

4. Rosa A. Eberly, *Citizen Critics: Literary Public Spheres* (Urbana, Illinois: University of Illinois Press, 2000), 20; David S. Kaufer and Kathleen M. Carley, *Communication at a Distance* (Hillsdale, NJ: Erlbaum, 1993), 11, 267.

5. Roger Cooter and Stephen Pumfrey, "Separate Spheres and Public Places: Reflections on the History of Science Popularization and Science in Popular Culture," *History of Science* 32 (1994): 242, 50.

6. Ibid., 237.

7. Charles Bazerman, *Shaping Written Knowledge* (Madison: University of Wisconsin Press, 1988), 133.

8. Clemens, "Memoranda: A Couple of Sad Experiences," 859.

9. Cooter and Pumfrey, "History of Science Popularization," 237n; Steven Shapin, "Science and the Public," in *The Companion to the History of Modern Science*, ed. R. C. Olby, et al. (New York: Routledge, 1990), 1001.

10. Carolyn R. Miller, "Genre as Social Action," *Quarterly Journal of Speech* 70, no. 2 (1984): 152–53.

11. Kaufer and Carley, *Communication at a Distance*, 254, 67.

12. Dwight Atkinson, *Scientific Discourse in Sociohistorical Context* (Mahwah, NJ: Erlbaum, 1999), 59.

13. Jane P. Tompkins, "The Reader in History," in *Reader-Response Criticism: From Formalism to Post-Structuralism*, ed. Jane P. Tompkins (Baltimore: Johns Hopkins University Press, 1980), 225.

14. James L. Machor, ed., *Readers in History: Nineteenth Century American Literature and the Contexts of Response* (Baltimore: Johns Hopkins University Press, 1993), x.

15. Kaufer and Carley, *Communication at a Distance*, 2.

16. Alan Prince and Paul Smolensky, *Optimality Theory: Constraint Interaction in Generative Grammar*, ms. ed. (New Brunswick and Boulder: Rutgers University and University of Colorado, 1993).

17. There could be other decisions in which they could compete with each other, say for instance in the reading of a plausible-sounding scientific discovery that nevertheless contained inconsistencies of reporting.

Chapter One.
A Brief Natural History of Hoaxing

1. The first hoax mentioned in the *OED* was the Great Stock Exchange Hoax of 1814. See explanation on p. 37.

2. Alexander Boese, *The Predictions of Isaac Bickerstaff* [webpage] (Boese, Alexander, 2002 [cited 10 February 2002]); available from http://www.museumof hoaxes.com/bickerstaff.html.

3. Ibid.

4. Dustin H. Griffin, *Satire: A Critical Reintroduction* (Lexington: University Press of Kentucky, 1994), 35, 38.

5. Ibid., 37–38.

6. Ibid., 95.

7. Jonathan Swift, *A Modest Proposal*, in The Norton Anthology of English Literature, vol. 1, 5th ed., edited by M. H. Abrams, (New York: W. W. Norton & Co., 1986); 2181–87.

8. Griffin, *Satire: A Critical Reintroduction*, 3.

9. Most of the media hoaxes in the eighteenth century were fake travel narratives, such as Defoe's wildly successful and controversial hoax autobiography *Robinson Crusoe*. This predilection for travel adventures likely fed the desire of readers to consume everything foreign during an era of imperialistic exploration and expansion by the English and other western European nations. Kaufer and Carley have added the suggestion that texts such as *Crusoe* and *Gulliver's Travels* fed a hunger for escapism created by the oppressive work schedules and landscapes of the Industrial Revolution in England. See Kaufer and Carley, *Communication at a Distance*, 71.

10. Boese, *The Predictions of Isaac Bickerstaff*; David D. Hall, "Readers and Reading in America: Historical and Critical Perspectives," *Proceedings of the American Antiquarian Society* 103, no. 2 (1993): 342.

11. Griffin, *Satire: A Critical Reintroduction*, 71.

12. See chapter 4 for a fuller exploration of the Socratic dialectic in Twain's hoaxing practices.

13. Jeanne Fahnestock and Marie Secor, "The Stases in Scientific and Literary Argument," *Written Communication* 5 (1998): 428–29.

14. Griffin, *Satire: A Critical Reintroduction*, 132.

15. Jonathan Swift, "A Modest Proposal," 1:2181.

16. Edgar Allan Poe, "A Predicament," in *The Complete Tales and Poems of Edgar Allan Poe* (New York: Modern Library, 1938), 328.

17. Edgar Allan Poe, "Hans Phaall—A Tale," *Southern Literary Messenger*, June 1835, 565.

18. Richard P. Benton, "Is Poe's 'The Assignation' a Hoax?" *Nineteenth-Century Fiction* 18, no. 1 (1963): 193.

19. Wilbur S. Shepperson, *Mirage-Land: Images of Nevada* (Reno: University of Nevada Press, 1992), 30.

20. Steven Mailloux, *Rhetorical Power* (Ithaca: Cornell University Press, 1989), 62.

21. Richard Adams Locke, *The Moon Hoax; or, a Discovery That That Moon Has a Vast Population of Human Beings*, The Gregg Press Science Fiction Series (Boston: Gregg, 1975), xxiii.

22. Clemens, "Memoranda: A Couple of Sad Experiences," 859.

23. H. Bruce Franklin, *Future Perfect: American Science Fiction of the Nineteenth Century*, revised ed. (New York: Oxford University Press, 1978), 93, 99.

24. Ibid., 96.

25. Robert V. Bruce, *The Launching of Modern American Science, 1846–1876*, first ed. (New York: Knopf, 1987), 25–27.

26. Frank Luther Mott, *A History of American Magazines: 1741–1850*, 5 vols., vol. 1 (Cambridge, Massachusetts: Harvard University Press, 1968), 342.

27. Frank Luther Mott, *A History of American Magazines: 1850–1865*, 5 vols., vol. 2 (Cambridge, Massachusetts: Harvard University Press, 1968), 157.

28. Ibid., 498.

29. Mott, *A History of American Magazines: 1741–1850*, 405.

30. Thomas O. Mabbott, ed., *Doings of Gotham in a Series of Letters by Edgar Allan Poe as Described to the Editor of the Columbia Spy; Together with Various Editorial Comments and Criticisms, by Poe* (Pottsville, Pennsylvania: Spannuth, 1929), 33.

31. Mott, *A History of American Magazines: 1741–1850*, 446.

32. Lee Rust Brown, *The Emerson Museum* (Cambridge, Massachusetts: Harvard University Press, 1997), 60.

33. C. P. Snow, *The Two Cultures: And a Second Look* (New York: New American Library, 1963), 30.

34. Kaufer and Carley, *Communication at a Distance*, 303.

35. Judith Yaross Lee, "Fossil Feuds: Popular Science and the Rhetoric of Vernacular Humor," *Essays in Arts and Sciences* 23 (1994): 3; Shapin, "Science and the Public," 1001.The term *elite* is often used to refer to the journals and societies forming during this period around a discourse that gradually became very difficult to follow for the readers of popular periodicals and the attendees of lyceum lectures. However, as Roger Cooter and Steven Pumfrey have pointed out, the use of *elite* implies a higher level of political as well as cultural power, and science's "relation with dominant culture is frequently problematic" (252). In this work I will use the term *specialized* to refer to the increasingly esoteric science journals proliferating in antebellum America and *professional* to refer to the scientific societies that became allied with university departments in the second half of the century.

36. Brown, *The Emerson Museum*, 138.

37. Mott, *A History of American Magazines: 1741–1850*, 446–47.

38. Stephen B. Katz, "Narration, Technical Communication, and Culture: *The Soul of a New Machine* as Narrative Romance," in *Constructing Rhetorical Education*, ed. Marie Secor and Davida Charney (Carbondale: Southern Illinois University, 1992), 384.

39. Dorothy Nelkin, *Selling Science: How the Press Covers Science and Technology*, revised ed. (New York: Freeman, 1995), 78.

40. Miles Orvell, *The Real Thing: Imitation and Authenticity in American Culture 1880–1940* (Chapel Hill: University of North Carolina Press, 1989), xv; Mark Seltzer, *Bodies and Machines* (New York: Routledge, 1992), 5.

41. Orvell, *The Real Thing: Imitation and Authenticity in American Culture 1880–1940*, xvii.

42. Lawrence I. Berkove, "Connecticut Yankee: Twain's Other Masterpiece," in *Making Mark Twain Work in the Classroom*, ed. James S. Leonard (Durham, NC: Duke University Press, 1999), 89; John Bryant, "Poe's Ape of Unreason: Humor, Ritual, and Culture," *Nineteenth-Century Literature* 51 (1996): 16.

Chapter Two. Method

1. Quoted in Gerald Graff, *Professing Literature: An Institutional History* (Chicago: University of Chicago Press, 1987), 69.

2. Ibid., 58.

3. Ibid., 68; R. H. Robins, *A Short History of Linguistics*, ed. R. H. Robins, Geoffrey Horrocks, and David Denison, 4th ed. (London: Longman, 1997), 207.

4. Graff, *Professing Literature: An Institutional History*, 73.

5. Robins, *A Short History of Linguistics*, 209.

6. Graff, *Professing Literature: An Institutional History*, 77.

7. Jacqueline Henkel, *The Language of Criticism: Linguistic Models and Literary Theory* (Ithaca, NY: Cornell University Press, 1996), 171.

8. Ibid., 1.

9. Jacqueline Henkel, "Linguistic Models and Recent Criticism: Transformational-Generative Grammar as Literary Metaphor," *PMLA* 105, no. 3 (1990): 456.

10. Henkel, *The Language of Criticism*, 14–18.

11. Ibid., 16.

12. Julie Tetel Andresen, *Linguistics in America, 1769–1924: A Critical History*, ed. Talbot J. Taylor, *Routledge History of Linguistic Thought Series* (New York: Routledge, 1990), 37.

13. Henkel, *The Language of Criticism*, 180–81.

14. Graff, *Professing Literature: An Institutional History*, 69.

15. Andresen, *Linguistics in America*, 33, 154.

16. Ibid., 156–57.

17. Ibid., 167.

18. Ibid., 148–49.

19. Geoffrey Sampson, *Schools of Linguistics* (Stanford: Stanford University Press, 1980), 104.

20. Robins, *A Short History of Linguistics*, 253.

21. Ibid.

22. Sampson, *Schools of Linguistics*, 127.

23. Andresen, *Linguistics in America*, 245; Henkel, *The Language of Criticism*, 2.

24. Robins, *A Short History of Linguistics*, 253.

25. Quoted in Graff, *Professing Literature*, 77.

26. Wayne C. Booth, *The Rhetoric of Fiction* (Chicago: University of Chicago Press, 1961), 67.

27. Henkel, *The Language of Criticism*, 11.

28. Andresen, *Linguistics in America*, 247.

Chapter Three. Poe's Hoaxing and the Construction of Readerships

1. Edgar Allan Poe, *Eureka: A Prose Poem* (New York: Prometheus Books, 1997), 1.

2. Kent P. Ljungquist, "'Valdemar' and the 'Frogpondians': The Aftermath of Poe's Boston Lyceum Appearance," in *Emersonian Circles: Essays in Honor of Joel Myerson*, ed. Wesley T. Mott (Rochester, NY: University of Rochester Press, 1997), 204.

3. Roland Barthes, "Textual Analysis of a Tale of Poe," in *On Signs*, ed. Marshall Blonsky (Baltimore: Johns Hopkins University Press, 1985), 94.

4. Arthur Hobson Quinn, *Edgar Allan Poe: A Critical Biography* (New York: Appleton-Century-Crofts, 1941), 71.

5. Ibid., 83–84.

6. Ibid., 109.

7. Ibid., 104.

8. Susan Booker Welsh, "Edgar Allan Poe and the Rhetoric of Science" (dissertation, Drew University, 1986), 90, 111.

9. Ibid., 231.

10. Poe, *Eureka: A Prose Poem*, 17.

11. Edgar Allan Poe, "Sonnet—to Science," in *The Raven and Other Poems* (New York: Wiley and Putnam, 1845), 771–72.

12. Lee, "Fossil Feuds: Popular Science and the Rhetoric of Vernacular Humor," 3.

13. Howard S. Miller, "The Political Economy of Science," in *Nineteenth-Century American Science: A Reappraisal*, ed. George H. Daniels (Evanston, IL: Northwestern University Press, 1972), 102.

14. John Richard Betts, "P. T. Barnum and the Popularization of Natural History," *Journal of the History of Ideas* 20 (1959): 357.

15. Neil Harris, *Humbug: The Art of P.T. Barnum* (Boston: Little, Brown, 1973), 73.

16. Jonathan Elmer, *Reading at the Social Limit: Affect, Mass Culture, and Edgar Allan Poe* (Stanford: Stanford University Press, 1995), 84.

17. Taylor Stoehr, *Hawthorne's Mad Scientists: Pseudoscience and Social Science in Nineteenth-Century Life and Letters* (Hamden, CT: Archon Books, 1978), 27.

18. Poe, *Eureka: A Prose Poem*, 44.

19. Edgar Allan Poe, "Von Kempelen and His Discovery," in *The Complete Tales and Poems of Edgar Allan Poe* (New York: Modern Library, 1938), 608.

20. Carolyn D. Hay, "A History of Science Writing in the United States and of the National Association of Science Writers" (dissertation, Northwestern University, 1970), 5, 9.

21. Bruce, *The Launching of Modern American Science, 1846–1876*, 72.

22. Ibid., 118.

23. Alexander Boese, "Dr. Egerton Yorrick Davis" *Museum of Hoaxes* http://www.museumofhoaxes.com/yorrick.html (accessed May 31, 2002).

24. Daniel Hoffman, *Poe Poe Poe Poe Poe Poe Poe*, f1st ed. (Garden City, N.J.: Doubleday, 1972), 86–87.

25. C. L. Barritt, "Why Are Not the Sciences Better Understood?" *Broadway Journal* 1, no. 8 (1845): 115.

26. Ibid.; 116. The negative implications of Barritt's criticism are clear: Barritt does not mention educating the working classes or women about science. With a few notable exceptions, like astronomer Maria Mitchell, women were not publicly recognized in the media as participants in science. While upper-class laymen were not participants either, exactly, they were at least educated to be spectators by the general science journals and other popular media. Women and the working classes were written into the story of American science in the decades before the Civil War only as patients and subjects. The advent of the penny press in 1835, which aimed below the upper-class belt, altered this situation slightly by assuming women, laborers, and immigrants as readers of its sensational scientific articles.

27. Edgar Allan Poe, "Ms Found in a Bottle," in *Complete Stories and Poems of Edgar Allan Poe* (New York: Doubleday, 1966), 148.

28. Hoffman, *Poe Poe Poe Poe Poe Poe Poe*, 185.

29. Marcy J. Dinius, "'Hans Phaall' the Great Moon Hoax, and Popular Science in the Antebellum American Imaginary" (paper presented at the Second Annual International Poe Conference, Baltimore, 3–6 October, 2002).

30. Edgar Allan Poe, "Note to 'Hans Phaall,'" in *The Moon Hoax*, ed. Ormond Seavey (Boston: Gregg, 1975), 69.

31. Poe, "Hans Phaall—A Tale," 565.

32. Dwight Thomas and David K. Jackson, *The Poe Log: A Documentary Life of Edgar Allan Poe 1809–1859* (Boston: Hall, 1987), 160.

33. Ibid., 161–62.

34. Edgar Allan Poe, *Letters of Edgar Allan Poe*, ed. John Ward Ostrom, 2 vols., vol. 1 (Cambridge, Massachusetts: Harvard University Press, 1948), 65.

35. Poe, "Hans Phaall—A Tale," 565.

36. Edgar Allan Poe, "The Literati of New York City: Richard Adams Locke," *E. A. Poe Society*, http://www.eapoe.org/works/misc/litratb6.htm (accessed January 4, 2001).

37. Fedler, *Media Hoaxes*, 68.

38. Nelkin, *Selling Science: How the Press Covers Science and Technology*, 85.

39. Sandra Harding claims that the industrial revolution had the opposite effect on the woman reader—that it kept her out of all serious centers of policy making because the industrial economy constructed her as less productive of market goods and therefore less valuable. Consequently, her education was not a priority. However, Mott and Harding are not necessarily arguing incommensurate points; it is feasible that women could be educated well enough to read the penny dailies but might still be excluded from the elite literary and scientific communities—therefore, the political communities.

40. Frank Luther Mott, *American Journalism: A History* (New York: Macmillan, 1962), 303.

41. Locke, *The Moon Hoax;* xxxi.

42. Ibid., 7.

43. Ibid., vii.

44. Ibid., xvi.

45. Ibid., xiii.

46. Karen S. H. Roggenkamp, *The Sun, the Moon, and Two Balloons: Edgar Allan Poe, Literary Hoaxes, and Penny-Press Journalism* http://roggenkamp.com/poe.html (accessed March 11, 2002).

47. These were Poe, Benson J. Lossing, Phillip Hone, William Griggs, and Harriet Martineau.

48. Harriet Martineau, *Retrospect of Western Travel* (New York, 1838), 22–23.

49. John Frederick William Herschel to Captain C[aldwell], 5 January 1836, Herschel Family Papers, Harry Ransom Center, The University of Texas at Austin.

50. Richard Adams Locke, *The Celebrated "Moon Story," Its Origin and Incidents; / with a Memoir of the Author, and an Appendix, Containing, I. An Authentic Description of the Moon; Ii. A New Theory of the Lunar Surface, in Relation to That of the Earth.*, ed. William N. Griggs (New York: Bunnell and Price, 1852), 30.

51. Poe, *Letters of Edgar Allan Poe*, 74.

52. Locke, *The Moon Hoax; or, a Discovery That That Moon Has a Vast Population of Human Beings*, 69.

53. Poe, The Literati of New York City: Richard Adams Locke.

54. Machor, ed., *Readers in History: Nineteenth Century American Literature and the Contexts of Response*, xxii.

55. Barthes, "Textual Analysis of a Tale of Poe," 86.

56. Locke, *The Moon Hoax*, 65.

57. Walter Kintsch, *Comprehension: A Paradigm for Cognition* (New York: Cambridge University Press, 1998), 94, 111.

58. The one slightly anomalous journal in this survey is the *American Journal of Science* (*AJS*), which was certainly considered a general science journal and was read more widely than the technical journals of the scientific societies. However, the *AJS* was not likely to have been read by the working-class target demographic of the penny dailies due to constraints of money, time, and education. It is included here to represent the general science reading material available to the 5 percent of Americans with a college education at this time, like Poe himself. Ellen Condliffe Lagemann, "Education to 1877," in *The Reader's Companion to American History*, ed. Eric Foner and John A. Garraty (Boston: Houghton Mifflin, 1991).

59. Totals and percentages for this category were adjusted as follows: the large number of articles in the educational biannual *Family Magazine* (53) was proportionally reduced in order to treat it statistically as a monthly, thus enabling a more equitable comparison with the other magazines and newspapers.

60. Bruce, *The Launching of Modern American Science, 1846–1876*, 62.

61. Charles Bazerman, "Reporting the Experiment: The Changing Account of Scientific Doings in the Philosophical Transactions of the Royal Society, 1665–1800," in *Landmark Essays on Rhetoric of Science: Case Studies*, ed. Randy Allen Harris (Mahwah, New Jersey: Hermagoras, 1997), 183; Jeanne Fahnestock, "Accomodating Science: The Rhetorical Life of Scientific Facts," *Written Communication* 3, no. 3 (1986): 291; John Swales, "Research Articles in English," in *Genre Analysis: English in Academic and Research Settings* (Cambridge, UK: Cambridge University Press, 1990), 118–19, 38, 40.

62. The only exceptions to this rule are found in a few of the *AJS* articles, which imitate the developing transactions-type experimental reports of this time period in restricting their claims to a specific hypothesis and its proof, a strategy that resists overgeneralization and sensationalism. Bazerman, "Reporting the Experiment: The Changing Account of Scientific Doings in the Philosophical Transactions of the Royal Society, 1665–1800," 174, 83.

63. Locke, *The Moon Hoax*, xxv; Poe, The Literati of New York City: Richard Adams Locke.

64. Poe, "Note to "Hans Phaall," 70.

65. Most of the criteria come from the version of the essay appearing in the *Columbia Spy*. The foreign, presentation, verisimilitude, and plausibility criteria are added in the version of the essay incorporated into Poe's "Literati" sketch of Locke.

66. Mabbott, ed., *Doings of Gotham*, 52–55.

67. In general this belief or "doxastic system" is coherent, disallowing contradictory beliefs about a single proposition. For a discussion of this sort of doxastic system in epistemology, see Keith Lehrer and Thomas D. Paxson Jr., "Knowledge, Undefeated Justified True Belief," in *Essays on Knowledge and Justification*, ed. George S. Pappas and Marshall Swain (Ithaca, NY: Cornell University Press, 1978). However, discourse researchers have found that discourse interpretation usually involves the acceptance of beliefs into the doxastic system that would be incompatible under a strict interpretation of the logical form of those beliefs. For example, we have no trouble accepting that penguins are birds, although they cannot fly, and flight counts as a defining feature of birds. In Alex Lascarides and Nicholas Asher, "Temporal Interpretation, Discourse Relations, and Commonsense Entailment," *Linguistics and Philosophy* 16 (1993), the authors argue that "defeasible" beliefs, beliefs that hold in general but that can be overridden by specific exceptions, allow more realistic modeling of the cognitive process of discourse interpretation and the update of belief systems that accompany it.

68. Kintsch, *Comprehension: A Paradigm for Cognition*, 95.

69. Locke, *The Moon Hoax*, xiii.

70. Poe, The Literati of New York City: Richard Adams Locke.

71. Leda Cosmides and John Tooby, "Consider the Source: The Evolution of Adaptations for Decoupling and Metarepresentation," in *Metarepresentation*, ed. Dan Sperber, *Vancouver Studies in Cognitive Science* (New York: Oxford University Press, 2000).

72. Franklin, *Future Perfect: American Science Fiction of the Nineteenth Century*, 94.

73. Poe, "Astounding News."

74. Mabbott, ed., *Doings of Gotham*, 33.

75. Thomas and Jackson, *The Poe Log: A Documentary Life of Edgar Allan Poe 1809–1859*, 461.

76. Mabbott, ed., *Doings of Gotham*, 33.

77. Ibid.

78. Thomas and Jackson, *The Poe Log*, 458.

79. Doris V. Falk, "Thomas Low Nichols, Poe, and the 'Balloon Hoax,'" *Poe Studies* 5, no. 2 (1972): 48.

80. Ibid., 49.

81. William Charvat, *The Profession of Authorship in America, 1800–1870: The Papers of William Charvat*, ed. Matthew J. Bruccoli (Columbus: Ohio State University Press, 1968), 28.

82. Theodore M. Porter, *Trust in Numbers: The Pursuit of Objectivity in Science and Public Life* (Princeton, NJ: Princeton University Press, 1995), 202–03.

83. Burton R. Pollin, "Poe's 'Von Kempelen and His Discovery': Sources and Significance," *Etudes Anglaises* 20 (1967): 14.

84. Franklin, *Future Perfect*, 94.

85. Locke, *The Celebrated "Moon Story,"* 21.

86. Poe, "Astounding News."

87. Edgar Allan Poe, "The Facts in the Case of M. Valdemar," *The American Review: A Whig Journal*, December 1845, 561.

88. Barthes, "Textual Analysis of a Tale of Poe," 87–89.

89. Steve Carter, "A Possible Source for 'the Facts in the Case of M. Valdemar,'" *Poe Studies* 12 (1979): 36.

90. Poe, *Letters of Edgar Allan Poe*, 319.

91. Elizabeth B. Barrett to Edgar Allan Poe, April 1846. In "Poe's Letters," *Edgar Allan Poe Society*, http://www.eapoe.org/misc/letters/t4604000.htm (accessed January 10, 2002).

92. Arch Ramsay to Edgar Allan Poe, 1846. In "Poe's Letters," *Edgar Allan Poe Society*, http://www.eapoe.org/misc/letters/t4611300.htm (accessed January 10, 2002).

93. Poe, *Letters of Edgar Allan Poe*, 337.

94. Barthes, "Textual Analysis of a Tale of Poe," 89.

95. Ibid., 92.

96. Ibid., 96.

97. Poe, "Von Kempelen and His Discovery," 606.

98. Ibid.

99. Ibid., 610.

100. Poe, *Letters of Edgar Allan Poe*, 433–34. Letter reprinted by permission of the publisher from *The Letters of Edgar Allan Poe, Volume II*, edited by John Ward Ostrom, pp. 433–34 (Cambridge, MA: Harvard University Press, copyright © 1948 by the President and Fellows of Harvard College).

101. Pollin, "Poe's 'Von Kempelen and His Discovery': Sources and Significance," 13.

102. Hoffman, *Poe Poe Poe Poe Poe Poe Poe*, 192.

103. Poe, "Von Kempelen and His Discovery," 605–06.

104. Ibid., 607.

105. Bruce, *The Launching of Modern American Science, 1846–1876*, 26.

106. Quinn, *Edgar Allan Poe, a Critical Biography*, 596.

107. Pollin, "Poe's 'Von Kempelen and His Discovery': Sources and Significance," 14.

108. Ibid., 19.

109. One of these tangents is a section that mocks George Eveleth, a medical student and regular correspondent of Poe, who tried to steal some of Poe's glory after the publication of *Eureka* by claiming to have already had and circulated some of the central ideas in it, according to Bernard Pollin (ibid., 17). Poe renames Eveleth "Kissam . . . or is it Mr. Quizzem." The passage reads in part, "It seems to me quite incredible that any man of common understanding could have discovered what Mr. Kissam says he did, and yet have subsequently acted so like a baby—so like an owl—as Mr. Kissam *admits* that he

did. By-the-way, who is Mr. Kissam? And is not the whole paragraph in the *Courier and Enquirer* a fabrication got up to 'make a talk'? It must be confessed that it has an amazingly moon-hoax-y air" (Poe, "Von Kempelen's Discovery," 606). The snide similes and chatty tone of this passage are very out of keeping even with the relatively serious language of the first paragraph of the story, reproduced in the text above. This language had to have thrown a wrench in readers' interpretive processes.

110. George Kennedy, "Literary Rhetoric," in *Classical Rhetoric and Its Christian and Secular Tradition from Ancient to Modern Times* (Chapel Hill: University of North Carolina Press, 1980), 111–12.

111. Franklin, *Future Perfect: American Science Fiction of the Nineteenth Century*, 319–20.

112. Fitz-James O'Brien, "How I Overcame My Gravity," *Harper's New Monthly Magazine* 28, no. 163 (1863): 779.

113. Poe, "Hans Phaall—A Tale," 565.

114. Benton, "Is Poe's "the Assignation" a Hoax?"; Bryant, "Ape"; Benjamin Franklin Fisher IV, "Poe's 'Tarr and Fether': Hoaxing in the Blackwood Mode," in *The Naiad Voice: Essays on Poe's Satiric Hoaxing*, ed. Dennis W. Eddings (Port Washington, NY: Associated Faculty, 1983); Richard Fusco, "Poe's Revisions of 'The Mystery of Marie Roget': A Hoax?" in *Poe at Work: Seven Textual Studies*, ed. Benjamin Franklin Fisher IV (Baltimore: Poe Society, 1978); G. R. Thompson, "Is Poe's "A Tale of the Ragged Mountains" a Hoax?" *Studies in Short Fiction* 6 (1969).

115. Thompson, "Is Poe's "A Tale of the Ragged Mountains" a Hoax?" 454.

116. Fisher IV, "Poe's 'Tarr and Fether': Hoaxing in the Blackwood Mode," 136.

117. Bryant, "Ape," 28.

118. Fusco, "Poe's Revisions of 'The Mystery of Marie Roget': A Hoax?" 92.

119. Edgar Allan Poe, "Murders in the Rue Morgue," in *Complete Stories and Poems of Edgar Allan Poe* (New York: Doubleday, 1938), 2.

120. Samuel Taylor Coleridge, *Biographia Literaria*, ed. J. Shawcross, vol. 2 (London: Oxford University Press, 1949), 6.

121. Marie-Louise Nickerson Matthew, "Forms of Hoax in the Tales of Edgar Allan Poe" (Columbia, 1975), 3.

122. It is interesting that Rolf Zwaan found in his study of the effects of genre on interpretation that participants who believed they were reading fiction (instead of a news story) reported appreciating the language of the story. Rolf Zwaan, "Effect of Genre Expectations on Text Comprehension," *Journal of Experimental Psychology: Learning, Memory, and Cognition* 20, no. 4 (1994). His finding jibes well with the reactions of contemporary readers of Locke's hoax who ranked entertainment highly in their interpretive decisions. While they did not care about the truth-value of the story, they reported enjoying its imaginative structure and language.

123. John Limon, *The Place of Fiction in the Time of Science* (New York: Cambridge University Press, 1990), 21.

124. Poe, *Eureka: A Prose Poem*, 33, 40–41.

125. Ibid., 5–6.

126. Locke, *The Moon Hoax; or, a Discovery That That Moon Has a Vast Population of Human Beings*, xxxiii.

127. Poe, *Eureka: A Prose Poem*, 8–9.

128. Ibid., 9–10.
129. Ibid., 10.
130. Ibid., 6.
131. Welsh, "Edgar Allan Poe and the Rhetoric of Science," 170.
132. Poe, *Eureka: A Prose Poem*, 16.
133. Welsh, "Edgar Allan Poe and the Rhetoric of Science," 170, 85–86.
134. The United States Exploring Expedition, begun in 1834, commanded by Captain Charles Wilkes. Along with Yale geologist James Dwight Dana, an illustrator and ethnographer attended the naval expedition to collect and record samples along the West Coast of the United States, the South Pacific, and Australia.
135. Mabbott, ed., *Doings of Gotham*, 50.
136. Poe, *Eureka: A Prose Poem*, 11.
137. Hoffman, *Poe Poe Poe Poe Poe Poe Poe*, 281.
138. Ibid.
139. Matthew, "Forms of Hoax in the Tales of Edgar Allan Poe," 73.
140. Edgar Allan Poe, "Fifty Suggestions," in *Edgar Allan Poe: Essays and Reviews*, ed. G. R. Thompson, *Library of America* (New York: Literary Classics of the United States, 1984), 1303.
141. Elmer, *Reading at the Social Limit: Affect, Mass Culture, and Edgar Allan Poe*, 30.
142. Poe, *Eureka: A Prose Poem*, 135.
143. Edgar Allan Poe, "Literati—Laughton Osborn," in *Edgar Allan Poe Collection*, Harry Ransom Center, University of Texas at Austin (Austin, Texas: 1846), 79.
144. Ljungquist, "'Valdemar' and the 'Frogpondians': The Aftermath of Poe's Boston Lyceum Appearance," 204.
145. James R. Guthrie, "Broken Codes, Broken Seals, and Stolen Poems in 'The Purloined Letter,'" *Edgar Allan Poe Review* 3, no. 2 (2002): 94.
146. Hoffman, *Poe Poe Poe Poe Poe Poe Poe*, 185.
147. Poe, *Eureka: A Prose Poem*, 6; Peter Swirski, *Between Literature and Science: Poe, Lem, and Explorations in Aesthetics, Cognitive Science, and Literary Knowledge* (Montreal: McGill-Queen's University Press, 2000), 28.
148. Edgar Allan Poe, "Letter to B———," in *Edgar Allan Poe: Essays and Reviews*, ed. G. R. Thompson, Library of America (New York: Literary Classics of the United States, 1984), 5.

Chapter Four. Mark Twain and the Social Mechanics of Laughter

1. Samuel Langhorne Clemens, *Mark Twain's Letters*, ed. Edgar Marquess Branch, Michael B. Frank, and Kenneth M. Sanderson, vol. 1, 1853–1866, The Mark Twain Papers (Berkeley: University of California Press, 1988), 323.
2. Samuel Langhorne Clemens, *The Autobiography of Mark Twain*, ed. Charles Neider (New York: HarperCollins, 2000), 40.
3. Ibid., 41.
4. Andrew Jay Hoffman, *Inventing Mark Twain: The Lives of Samuel Langhorne Clemens* (New York: Morrow, 1997), 15.

5. Samuel Langhorne Clemens, *Mark Twain of the Enterprise: Newspaper Articles & Other Documents 1862–1864*, ed. Henry Nash Smith (Berkeley: University of California Press, 1957), 136–37.

6. John Lauber, *The Making of Mark Twain: A Biography* (New York: American Heritage, 1985), 46–47.

7. Everett H. Emerson, *Mark Twain: A Literary Life* (Philadelphia: University of Pennsylvania Press, 2000), 8.

8. Clemens, *The Autobiography of Mark Twain*, 128.

9. Emerson, *Mark Twain: A Literary Life*, 4.

10. The one exception to this rule of which I am aware was an 1899 hoax by Denver newsmen about the Great Wall of China being dismantled. However, China, too, was a sort of American frontier as it was the focus of foreign trade efforts after the Civil War (cf. the "trade dollar" controversy in chapter 5).

11. Walter Blair and Hamlin Hill, *America's Humor: From Poor Richard to Doonesbury* (New York: Oxford University Press, 1978), 260.

12. Pascal Covici Jr., *Mark Twain's Humor: The Image of a World* (Dallas: Southern Methodist University Press, 1962), 8. "Author" is slightly misleading because these forms were primarily oral and were only recorded by authors after they had been in circulation for decades.

13. James M. Cox, *Mark Twain: The Fate of Humor* (Princeton, NJ: Princeton University Press, 1966), 98.

14. Covici, *Mark Twain's Humor: The Image of a World*, 731.

15. Samuel Langhorne Clemens, *Roughing It* (Hartford, CT: American, 1891), 61–66.

16. Covici, *Mark Twain's Humor: The Image of a World*, 27; Constance Rourke, *American Humor* (New York: Harcourt Brace, 1931), 2:4.

17. Cox, *Mark Twain: The Fate of Humor*, 15; Judith Yaross Lee, "(Pseudo-) Scientific Humor," in *American Literature and Science*, ed. Robert J. Scholnick (Lexington: University Press of Kentucky, 1992), 142.

18. Clemens, *Roughing It*, 241–47.

19. James W. Cook, *The Arts of Deception: Playing with Fraud in the Age of Barnum* (Cambridge, MA: Harvard University Press, 2001), 26–27.

20. William Wright, *Dives and Lazarus*, ed. Lawrence I. Berkove (Ann Arbor, MI: Ardis, 1988), 22.

21. Covici, *Mark Twain's Humor: The Image of a World*, 146.

22. Rourke, *American Humor*, 26

23. Blair and Hill, *America's Humor: From Poor Richard to Doonesbury*, 228.

24. Albert Bigelow Paine, *Mark Twain: A Biography* (New York: Harper & Brothers, 1912), 512.

25. Clemens, *The Autobiography of Mark Twain*, 127.

26. Hyatt Howe Waggoner, "Science in the Thought of Mark Twain," *American Literature* 8, no. 4 (1937): 362.

27. David Ketterer, "The 'Science Fiction' of Mark Twain," *Mosaic—A Journal for the Interdisciplinary Study of Literature* 16, no. 4 (1983): 69; Waggoner, "Science in the Thought of Mark Twain," 361.

28. Waggoner, "Science in the Thought of Mark Twain," 359.

29. Lee, "(Pseudo-) Scientific Humor," 129.

30. Ibid., 132.

31. Waggoner, "Science in the Thought of Mark Twain," 359.

32. Bruce, *The Launching of Modern American Science, 1846–1876*, 150–55.

33. Leo Marx, *The Machine in the Garden: Technology and the Pastoral Ideal in America* (New York: Oxford University Press, 1964; reprint, 2000), 321.

34. Bruce, *The Launching of Modern American Science, 1846–1876*, 276.

35. The one exception was the field of medicine, which seemed the least affected by the Civil War and which was perhaps even advanced by the service of field doctors and by the abolition of slavery. Darlene Clark Hine describes the tangible gains for blacks in the South due to the medical schools established for blacks after the Civil War. In the twenty-five years following the war, 115 black women were certified as physicians, and by 1890 there were 909 black male doctors (210–11). See Hine, "Colaborers in the Work of the Lord: Nineteenth-Century Black Women Physicians," in *Racial Economy of Science*, ed. Sandra Harding (Boulder, CO: Netlibrary, Inc., 1999), 526.

36. Samuel Langhorne Clemens, "A Private History of a Campaign That Failed," in *The American Claimant and Other Stories and Sketches* (New York: Harper & Brothers, 1923), 276–79.

37. H. Bruce Franklin, "Traveling in Time with Mark Twain," in *American Literature and Science*, ed. Robert J. Scholnick (Lexington: University Press of Kentucky, 1992), 166.

38. Marx, *The Machine in the Garden*, 330.

39. Cynthia E. Russett, *Darwin in America* (San Francisco: Freeman, 1976), 11.

40. Ibid., 10.

41. Ibid., 18.

42. Ibid., 3.

43. James D. Wilson, "'The Monumental Sarcasm of the Ages': Science and Pseudoscience in the Thought of Mark Twain," *South Atlantic Bulletin: A Quarterly Journal Devoted to Research and Teaching in the Modern Languages and Literatures* 40, no. 2 (1975): 79.

44. Berkove, "Connecticut Yankee: Twain's Other Masterpiece," 108.

45. Samuel Langhorne Clemens, *Early Tales and Sketches*, ed. Edgar Marquess Branch, Robert H. Hirst, and Harriet Elinor Smith, vol. 1, 1851–1864, *Works of Mark Twain* (Berkeley, California: University of California Press, 1979), 158.

46. Charles Bazerman, "Physicists Reading Physics," *Written Communication* 2 (1985); Davida Charney, "A Study in Rhetorical Reading: How Evolutionists Read 'The Spandrels of San Marco,'" in *Understanding Scientific Prose*, ed. Jack Selzer (Madison: University of Wisconsin Press, 1993).

47. Bertrand Gervais, "Reading Tensions: Of Sterne, Klee, and the Secret Police," *New Literary History* 26 (1995): 160.

48. There is another consequence of comprehension losing this competition, and that is the deactivation of internal coherence and detail as working expectations, because a reader's ability to judge the internal consistency of a story is severely hampered by skimming and missing the details, which may or may not add up.

49. Lee, "(Pseudo-) Scientific Humor," 129; Mott, *A History of American Magazines: 1850–1865*, 298.

50. Clemens, *Mark Twain of the Enterprise: Newspaper Articles & Other Documents 1862–1864*, 7.

51. Ibid., 27.

52. Mott, *A History of American Magazines: 1850–1865*, 289.

53. Paula Mitchell Marks, *Precious Dust: The American Gold Rush Era: 1848–1900* (New York: Morrow, 1994), 171.

54. Granville Stuart, *The Montana Frontier 1852–1864* (Lincoln: University of Nebraska Press, 1925; reprint, 1977), 270.

55. Marks, *Precious Dust: The American Gold Rush Era: 1848–1900*, 40.

56. Ibid., 198.

57. Lest we think we have outgrown these immature reading habits, David Perlman reminds us that mercenary urgency is still a major contributor to instances of hasty or inaccurate reporting of science news in the United States. In David Perlman, "Science and the Mass Media," in *Science and Its Public: The Changing Relationship*, ed. G. Holton and W. Blanpied (Boston: Dordrecht, 1976), 253.

58. James L. Tyson, *Diary of a Physician in California* (Oakland, CA: Biobooks, 1955), 75.

59. Quoted in Shepperson, *Mirage-Land: Images of Nevada*, 34–35.

60. William Dennison Bickham, *A Buckeye in the Land of Gold: The Letters and Journals of William Dennison Bickham*, ed. Randall E. Ham (Spokane, WA: Clark, 1996), 71.

61. Ibid., entry for 18 January 1851.

62. Clemens, "Memoranda: A Couple of Sad Experiences," 858–59.

63. Ibid.; 859–60.

64. DeLancey Ferguson, "The Petrified Truth," *Colophon II* (n.s.) 2 (1937): 193.

65. Clemens, "Memoranda: A Couple of Sad Experiences," 859.

66. Ibid.

67. Ibid.

68. Clemens, *Mark Twain's Letters*, 242.

69. Clemens, *Early Tales and Sketches*, 157.

70. William Wright, *Reporting with Mark Twain* (San Francisco: California Publishing, 1893), 72.

71. Clemens, "Memoranda: A Couple of Sad Experiences," 861.

72. Ibid., 859.

73. Ibid., 861.

74. Covici, *Mark Twain's Humor: The Image of a World*, 150.

75. Ibid., 150–51.

76. Clemens, "Memoranda: A Couple of Sad Experiences," 861.

77. Ibid., 858.

78. Carol Berkenkotter and Thomas Huckin, "News Value," in *Genre Knowledge in Disciplinary Communication* (Mahwah, NJ: Erlbaum, 1995), 31; Teun Van Dijk, "News Schemata," in *Studying Writing: Linguistic Approaches*, ed. S. Greenbaum and Cooper (Beverly Hills: Sage, 1986).

79. Clemens, "Memoranda: A Couple of Sad Experiences," 858.

80. Wright, *Reporting with Mark Twain*, 172.

81. Ibid., 171.

82. Clemens, *Mark Twain of the Enterprise: Newspaper Articles & Other Documents 1862–1864*, 9.

83. Ibid.

84. Totals for *AJS* (semiannual) and *Scientific American* (weekly) have been adjusted to be comparable with the dailies.

85. Lee, "(Pseudo-) Scientific Humor," 141.

86. Joan Belcourt Ross, "Mark Twain and the Hoax" (dissertation, Purdue, 1974), 1, 3.

87. Berkove, "Connecticut Yankee: Twain's Other Masterpiece," 89.

88. Covici, *Mark Twain's Humor: The Image of a World*, 154.

89. Ibid., 256.

90. Samuel Langhorne Clemens, "Double-Barreled Detective Story," *Harper's New Monthly*, January 1902, 264.

91. Covici, *Mark Twain's Humor: The Image of a World*, 144.

92. Waggoner, "Science in the Thought of Mark Twain," 357.

93. Shelley Fisher Fishkin, *Lighting out for the Territory: Reflections on Mark Twain and American Culture* (New York: Oxford University Press, 1996), 179–80.

94. Berkove, "Connecticut Yankee: Twain's Other Masterpiece," 90.

95. Ross, "Mark Twain and the Hoax."

96. Samuel Langhorne Clemens, "3,000 Years among the Microbes," in *Which Was the Dream?* ed. John S. Tuckey (Berkeley: University of California Press, 1968), 454.

97. Ibid.

98. Ibid., 487.

99. Ibid., 551–53.

100. Beverly A. Hume, "Twain's Satire on Scientists: Three Thousand Years among the Microbes," *Essays in Arts and Sciences* 16 (1997): 80.

101. Franklin, "Traveling in Time with Mark Twain," 170.

102. Samuel Langhorne Clemens, "The American Claimant," in *The American Claimant and Other Stories and Sketches* (New York: Harper & Brothers, 1923), 83.

103. Ibid.

104. Ibid., 80.

105. Wilson, "'The Monumental Sarcasm of the Ages': Science and Pseudoscience in the Thought of Mark Twain," 81.

106. Waggoner, "Science in the Thought of Mark Twain," 367.

107. Wilson, "'The Monumental Sarcasm of the Ages': Science and Pseudoscience in the Thought of Mark Twain," 81.

108. Ross, "Mark Twain and the Hoax," 189.

109. Cox, *Mark Twain: The Fate of Humor*, 146.

110. Ross, "Mark Twain and the Hoax," 9

111. For a discussion of negative and positive face as they bear on politeness and other pragmatic aspects of human communication, see P. Brown, and S. C. Levinson, *Politeness: Some Universals in Language Usage.* Studies in Interactional Sociolinguistics 4 (Cambridge: Cambridge University Press, 1987).

112. Rourke, *American Humor*, 26.

113. Kaufer and Carley, *Communication at a Distance*, 411, 16.

114. Voltaire, *Candide*, trans. Shane Weller (New York: Dover, 1993), 139.

115. Clemens, "3,000 Years among the Microbes," 496.

116. Ross, "Mark Twain and the Hoax," 189.

117. Covici, *Mark Twain's Humor: The Image of a World*, 159.

118. Rourke, *American Humor*, 63.

119. Ibid., 72.

120. Samuel Langhorne Clemens, "To the Person Sitting in Darkness," in *Mark Twain's Weapons of Satire: Anti-Imperialist Writings on the Philippine-American War*, ed. Jim Zwick (Syracuse: Syracuse University Press, 1992), 2.

121. Jim Zwick, Mark Twain and Imperialism in Anti-Imperialism in the United States, 1898–1935 (ed. Jim, Zwick, 2002 [cited 18 August 2002]); available from http://www.boondocksnet.com/ai/twain/index.html.

122. Warwick Wadlington, *The Confidence Game in American Literature* (Princeton: Princeton University Press, 1975), 195.

123. In Wilson, "'The Monumental Sarcasm of the Ages': Science and Pseudoscience in the Thought of Mark Twain."

Chapter Five. The Hoaxes of Dan De Quille

1. William Wright, "Quille Drops—Those Pre-Historic Tracks," *Territorial Enterprise*, 23 November 1885.

2. Richard A. Dwyer and Richard E. Lingenfelter, *Dan De Quille, the Washoe Giant: A Biography and Anthology*, Western Literature Series (Reno: University of Nevada Press, 1990); William Wright, *Dives and Lazarus*, ed. Lawrence I. Berkove (Ann Arbor, MI: Ardis, 1988).

3. William Wright, *The Big Bonanza*, ed. Oscar Lewis (New York: Apollo Editions, 1947), viii.

4. Wright, *Dives and Lazarus*, 30.

5. C. C. Goodwin, "Dan and His Quills," *Territorial Enterprise*, 3 February 1876, 2.

6. Wright, *Dives and Lazarus*, 31.

7. All references to the De Quille papers in the body of the text are given with the appropriate container numbers. The full citation of the papers is found in the Bibliography.

8. Fldr. 42.

9. Wright, *Reporting with Mark Twain*, 171.

10. Ibid.

11. Ctn. 2.

12. Wright, *Dives and Lazarus*, 16.

13. Wells Drury, *An Editor on the Comstock Lode* (Reno: University of Nevada Press, 1984), 211.

14. Wright, *The Big Bonanza*, 71.

15. Ctn. 1, fldr. 50.

16. Ctn. 1, fldr. 8.

17. C. Grant Loomis, "The Tall Tales of Dan De Quille," *California Folklore Quarterly* 5, no. 1 (1946): 30.

18. C. C. Goodwin, "Dan De Quille," in *As I Remember Them* (Salt Lake City, UT: Salt Lake Commercial Club, 1913), 214.

19. Lee, "(Pseudo-) Scientific Humor," 144.

20. Loomis, "The Tall Tales of Dan De Quille," 29.

21. Ibid., 37.

22. Lee, "(Pseudo-) Scientific Humor," 142.

23. William Wright, "The Wonder of the Age: A Silver Man," *Golden Era*, 5 February 1865, 3.

24. Ibid., 4.

25. Harris, *Humbug: The Art of P.T. Barnum*, 72.

26. George W. Chesley, "Dictation from George W. Chesley," in *Hubert Howe Bancroft Collection* (BANC MSS C-D 810:72, Bancroft Library, University of California, Berkeley, 1888).

27. Wright, "The Wonder of the Age: A Silver Man," 3.

28. The obvious shortage of landforms named for male sex organs, in spite of the abundance of glaringly suggestive opportunities in the West, raises an interesting connection between strange landscapes and gendered others who must be feared and dominated.

29. Wright, "The Wonder of the Age: A Silver Man," 3.

30. Ibid., 4.

31. Ctn. 1, fldr. 120.

32. William Wright, letter, 23 August 1874.

33. William Wright, "Sad Fate of a Nevada Inventor," *Scientific American*, 25 July 1874, 51.

34. Goodwin, "Dan De Quille," 216.

35. Ferguson, "The Petrified Truth," 193.

36. Ctn. 1, fldr. 120.

37. Ctn. 1, fldr. 120.

38. Ibid.

39. Ibid.

40. William Wright, "Traveling Stones," *Territorial Enterprise*, 26 October 1867, 3.

41. Ctn. 2, fldr. 98.

42. Ctn. 2, fldr. 3.

43. Drury, *An Editor on the Comstock Lode*, 212–13.

44. Dwyer and Lingenfelter, *Dan De Quille, the Washoe Giant*, 22.

45. Harris, *Humbug: The Art of P. T. Barnum*, 82.

46. See http://www.angelfire.com/ca7/xato/newtilt.html for some pictures of the traveling stones.

47. Ctn. 1, fldr. 120.

48. Goodwin, "Dan De Quille," 215.

49. Ctn. 1.

50. Ctn. 1, fldr. 120.

51. Ctn. 2, scrapbk. 2. "Trade dollars" were a controversial minting of heavy silver dollars by the U.S. Mint in 1873 in order to shoulder the standard Mexican peso out of currency in the Far East. According to numismatic historian Anthony Vigliotta, some of the "trade dollars" trickled back to the States, which they were never meant to do and, coupled with a drop in silver stock prices in 1876, created an unfortunate surplus of silver

currency, especially on the West Coast. This led to employers abusing their employees by paying them the undervalued trade dollars, which were refused by many merchants as indicated in the Union's response to the Eyeless Fish hoax. Overall, silver and "free silver" issues in the West are complicated economic and political problems that cannot be done justice in a short space. A more detailed treatment will follow in a discussion of De Quille's role in the free silver movement.

52. Ibid.
53. Ibid.
54. Ctn. 2, scrapbk. 5.
55. Ctn. 2, scrapbk. 2.
56. Ctn. 1, scrapbk. 120.
57. Box 3.
58. Wright, *Dives and Lazarus*, 22.
59. Kaufer and Carley, *Communication at a Distance*, 34, 67.
60. Ctn. 1, fldr. 8.
61. Wright, *Dives and Lazarus*, 21.
62. Drury, *An Editor on the Comstock Lode*, 213–14.
63. Wright, *Dives and Lazarus*, 39.
64. Ctn. 1, fldr. 39.
65. Ctn. 2.
66. Box 3, fldr. "miscellany."
67. Wright, "The Wonder of the Age: A Silver Man," 3.
68. Wright, *Dives and Lazarus*, 35.
69. Ibid., 32–33.
70. Ibid.
71. Dwyer and Lingenfelter, *Dan De Quille, the Washoe Giant*, 22.

Chapter Six. The Mechanics of Hoaxing

1. John Muir, "Nevada's Dead Towns," in *Steep Trails*, ed. William Frederic Badè (Boston: Houghton Mifflin, 1918), 195.

2. Ibid., 196–203.

3. Karlyn Kohrs Campbell, "Consciousness-Raising: Linking Theory, Criticism, and Practice," *Rhetoric Society Quarterly* (2002): 49.

4. Why did women not write hoaxes? A purely statistical answer is that there were simply not many women reporters for popular media in the mid-1800s (if any) and no women popular science writers (to my knowledge), so women simply lacked the opportunity for large-scale scientific media hoaxing. However, this is not a satisfactory explanation. A naïve hypothesis might hold that hoaxing, with its agonistic emphases on criticism and dominance, was gendered male rather than female at this time, though this generalization begs scrutiny and corroborating research. As a potential counterpoint to this argument, women did read and publicly respond to hoaxes (cf. Elizabeth B. Browning's letter to Poe about M. Valdemar and Harriet Martineau's commentary on Locke's Moon Hoax). A more suggestive reading of the gendering of hoaxes keys off Blair and Hill's classification of hoaxing as "subversive." In Jane Tompkins, *Sensational Designs: The Cultural Work of American Fiction, 1790–1860* (New York: Oxford University Press, 1985). Jane Tompkins has made an excellent argument that women writers

pioneered their own highly effective "subversive" mode of social criticism and subversive cultural work that fell within the bounds of respectable femininity in antebellum America—the sentimental novel. Chapter 5 of *Sensational Designs* frames the issues of gender and power associated with this literature through the lens of Harriet Beecher Stowe's *Uncle Tom's Cabin*.

5. De Quille obviously derived satisfaction from fooling scientists Baird and Cope, but his personal papers leave us no evidence of his use of the hoax to satisfy a personal vendetta.

6. Paul Collins, *Banvard's Folly: Thirteen Tales of People Who Didn't Change the World* (New York: Picador USA, 2001), 63.

7. Orvell, *The Real Thing*, xvii.

8. Marx, *The Machine in the Garden*, 236.

9. Clemens, *Mark Twain's Letters*, 323.

10. Ralph Waldo Emerson, *Early Poems of Ralph Waldo Emerson* (New York: Crowell, 1899).

11. Henry Adams, *Letters of Henry Adams*, ed. Worthington Chauncey Ford, vol. 1 (New York: Houghton Mifflin, 1930), 290.

12. Thomas and Jackson, *The Poe Log: A Documentary Life of Edgar Allan Poe 1809–1859*, 305.

13. Ibid., 514.

14. Elizabeth Tebeaux, *The Emergence of a Tradition: Technical Writing in the English Renaissance, 1475–1640*, Baywood's Technical Communications Series (Amityville, NY: Baywood, 1997), 84.

15. Cecilia Tichi, *Shifting Gears: Technology, Literacy, and Culture in Modernist America* (Chapel Hill: University of North Carolina Press, 1987), xii.

16. Bruno Latour, *We Have Never Been Modern*, trans. Catherine Porter (Cambridge, MA: Harvard University Press, 1993), 12.

17. Barry Barnes, *Scientific Knowledge and Sociological Theory* (Boston: Routledge & Kegan Paul, 1974), 2.

Conclusion: The Sokal Hoax

1. Alan Sokal, "Revelation: A Physicist Experiments with Cultural Studies," in *The Sokal Hoax: The Sham That Shook the Academy*, ed. editors of *Lingua Franca* (Lincoln: University of Nebraska Press, 2000), 49.

2. The only exception to the random sampling was as follows: if a particular issue of *Social Text* contained a section called "Science Studies," then the test article was randomly selected from this section. The goal was to recreate Sokal's reading practices, and he claimed in his *apologia* for his hoax to have been motivated to write it by purported abuses of scientific principles by science studies writers.

3. A common conclusion in cultural studies analyses. I am indebted to Barry Sarchett for pointing out this trend in the data.

4. Alan Sokal, "Transgressing the Boundaries: Towards a Transformative Hermeneutics of Quantum Gravity," *Social Text* 46/47 (1996): 226.

5. Ibid., 229.

6. R. Martin, "Resurfacing Socialism: Resisting the Appeals of Tribalism and Localism," *Social Text* 44, no. Autumn–Winter (1995): 102.

7. Sokal, "Transgressing the Boundaries: Towards a Transformative Hermeneutics of Quantum Gravity," 218.

8. Ibid., 217.

9. Ibid., 218.

10. Ibid., 218–19.

11. Bruce Robbins and Andrew Ross, "Response: Mystery Science Theater," in *The Sokal Hoax: The Sham That Shook the Academy*, ed. editors of *Lingua Franca* (Lincoln: University of Nebraska Press, 2000), 55.

12. Sokal, "Transgressing the Boundaries: Towards a Transformative Hermeneutics of Quantum Gravity," 225.

13. In "Moving beyond the Moment," Paul, Kendall, and Charney (2001) argue that historical evaluations of the "success" of certain rhetorical strategies, in the absence of evidence of contemporary responses by readers, run the risk of assuming the very success that they are trying to demonstrate.

14. Alan Sokal, "Reply to Michael Bérubé," in *The Sokal Hoax: The Sham That Shook the Academy*, ed. editors of *Lingua Franca* (Lincoln: University of Nebraska Press, 2000), 144.

15. Robbins and Ross, "Response: Mystery Science Theater," 55.

16. Alan Sokal, "A Physicist Experiments with Cultural Studies," *Lingua Franca* May/June (1996): 64.

17. Alan Sokal, *A Plea for Reason, Evidence, and Logic* (paper presented at the Forum on the Sokal Hoax, New York University, 30 October 1996), http://www.physics.nyu.edu/~as2/nyu_forum.html (accessed September 15, 2003).

18. I am indebted to Davida Charney for this insight. For a contrasting case of peer reaction to scientists creating controversy in their own field, see Charney's "A Study in Rhetorical Reading: How Evolutionists Read 'The Spandrels of San Marco.'"

19. Bruno Latour, "Is There Science after the Cold War?" in *The Sokal Hoax: The Sham That Shook the Academy*, ed. editors of *Lingua Franca*_(Lincoln: University of Nebraska Press, 2000), 124.

20. Robbins and Ross, "Response: Mystery Science Theater," 58.

21. Editors of *Lingua Franca*, ed., *The Sokal Hoax: The Sham That Shook the Academy* (Lincoln: University of Nebraska Press, 2000), 6.

22. Kaufer and Carley, *Communication at a Distance*, 2

23. Peter Osborne, "Friendly Fire: The Hoaxing of *Social Text*," in *The Sokal Hoax: The Sham That Shook the Academy*, ed. editors of *Lingua Franca* (Lincoln: University of Nebraska Press, 2000), 196.

24. An interesting example is hoaxer Joey Skaggs, who through clever press releases has gotten TV and print media to broadcast several of his hoaxes over the last thirty years, including a "Fat Squad" that bullies its clients into thinness, and "Solomon," an artificial-intelligence jury that Skaggs "announced" in the wake of the O. J. Simpson trial; it was reported by CNN and CBS radio news, among other media.

25. In Kaufer and Carley, *Communication at a Distance*, the authors find, against Walter Ong's hypothesis that new media gradually kill old media, that in fact combining multiple media—like print and the internet—augments the speed and range of communications (6). This finding may explain why Sokal quickly shifted the central action of his hoax to the Internet: he was both responding to the hoax's proliferation to electronic media and facilitating it in order to reach as wide an audience as possible.

26. Osborne, "Friendly Fire: The Hoaxing of *Social Text*," 197.

27. Sokal, "A Physicist Experiments with Cultural Studies," 64.

28. Andrew Ross, Stanley Aronowitz, and Bruno Latour, among others, strenuously objected to this claim, arguing that no sane cultural critic thinks there is no such thing as reality. They corrected Sokal's misapprehension by saying that what is at issue in cultural studies of science is our inability to separate fact from value when claims about reality become public and therefore inescapably rhetorical.

Glossary

The following terms and symbols are essential for understanding optimality theory as it is adapted in this project:

*: indicates one violation of a particular reading expectation

!: an exclamation mark placed after a violation "*" indicates that that violation is fatal, i.e., that candidate interpretation is now disqualified from the competition to be the optimal interpretation.

candidate: a possible interpretation of a text, i.e., "true" or "false" for potential science hoaxes. The interpretation that best survives the "filter" of the reader's preconceptions about science and science news is identified as the *optimal candidate.* Occasionally the reader will have insufficient information to select an optimal candidate or will value two conflicting preconceptions equally.

constraint: the term in optimality theory for a rule that governs a particular decision matrix. Constraints are stated as propositions. For science newsreading constraints, see the definition of "expectation."

dominate: to be ranked higher than in importance or value. Domination is determined on the basis of readers' decisions in the face of competing expectations. For example, if a reader chooses to believe a sensational news item regardless of the fact that it is riddled with scientific errors, sensation crucially dominates internal coherence for that reader.

expectation: a preconception that influences interpretation while reading. Expectations are stated as propositions that are either met or violated during a particular reading experience of a particular text. The expectations used in this book are expectations of genre (science news) and ethnoscience (popular cultural knowledge about science):

Authority: The author or authority figure's previous reputation holds.

Comprehension: It is optimal to comprehend everything written in a story.

Corroboration: If a science news story is corroborated by several news sources, it is probably true. (Local western readers interpreted "news sources" as eyewitnesses, not texts.)

Entertainment: Reading of popsci. articles is for entertainment, not truth, value. (Deactivated in truth-value games.)

Foreign: Anything foreign is good and probably true.

Internal Coherence: If the claims made by a story are logically consistent, it is probably true.

Medium: The previous reputation of the medium holds.

News: It is optimal to read news nonlinearly, skipping from the summary to main events and details of main events sections first and only reading background and supporting material as time and interest allow.

Novelty: New discoveries are highly valued and probably true.

Plausibility: If it seems like it could happen, it probably did.

Popsci: Stories that sound like true science reports probably are. (Subexpectations within this category are as follows for 1835–1850; they change to conform to the news schema above during 1862–1880:)

Problem/Solution: Popular science reports are usually structured on a problem/solution topos. (Related to internal coherence.)

Long: Longer popular science articles are often given in installments.

Decoration: Popular science reports will often be decorated with bold headlines and woodcuts.

Mystery: Popular science reports often have a "mystery" opening signaled by words such as *wonders*.

Ignorance/Wisdom: Popular science reports, after the opening, often employ an ignorance/wisdom antithesis to segue from problem to solution. (Related to authority and foreign.)

Detail: Popular science articles will often have a lot of technical detail.

Analogy: The details in a popular science article will often be explained with analogy to well-known phenomena.

Use: Popular science articles often finish with an evaluation of the benefit, physical or metaphysical, of the scientific principle/phenomenon.

Progression: It is optimal to read a story as fast as possible. Competes with comprehension.

Sensation: (1835–1850) Sensational elements in a story have a high literary and truth-value. (1862–1880) Sensational science news stories are probably not true.

ranking: a crucial ordering of expectations based on importance to the reader. See *dominate*. Ranking is indicated graphically by solid vertical lines separating levels of rank; everything to the left of a solid line dominates everything to the right of it. Equality of rank is indicated by dotted vertical lines between expectations that do not compete with each other and therefore are at the same level of ranking in the reader's value system.

Bibliography

Adams, Henry. *Letters of Henry Adams.* Edited by Worthington Chauncey Ford. Vol. 1. New York: Houghton Mifflin, 1930.

———, ed. *The Education of Henry Adams.* New York: Houghton Mifflin, 1918.

Andresen, Julie Tetel. *Linguistics in America, 1769–1924: A Critical History.* Edited by Talbot J. Taylor, Routledge History of Linguistic Thought Series. New York: Routledge, 1990.

Atkinson, Dwight. *Scientific Discourse in Sociohistorical Context.* Mahwah, NJ: Erlbaum, 1999.

Bancroft, Hubert Howe, Collection. BANC MSS C-D 810:72, Bancroft Library, University of California, Berkeley.

Barbosa, Pilar et al., ed. *Is the Best Good Enough? Optimality and Competition in Syntax.* Cambridge, MA: MIT Press, 1998.

Barnes, Barry. *Scientific Knowledge and Sociological Theory.* Boston: Routledge & Kegan Paul, 1974.

Barritt, C. L. "Why Are Not the Sciences Better Understood?" *Broadway Journal* 1, no. 8 (1845): 115–17.

Barthes, Roland. "Textual Analysis of a Tale of Poe." In *On Signs,* edited by Marshall Blonsky. Baltimore: Johns Hopkins University Press, 1985.

Bazerman, Charles. *The Languages of Edison's Light.* online ed. Cambridge, MA: MIT Press, 1999, http://www.netlibrary.com/ebook_info.asp?productid=10717 (accessed January 11, 2003).

———. "Physicists Reading Physics." *Written Communication* 2 (1985): 3–24.

———. "Reporting the Experiment: The Changing Account of Scientific Doings in the Philosophical Transactions of the Royal Society, 1665–1800." In *Landmark Essays on Rhetoric of Science: Case Studies,* edited by Randy Allen Harris, 241. Mahwah, NJ: Hermagoras, 1997.

———. *Shaping Written Knowledge.* Madison: University of Wisconsin Press, 1988.

Benton, Richard P. "Is Poe's "The Assignation" a Hoax?" *Nineteenth-Century Fiction* 18, no. 1 (1963): 193–97.

Berkenkotter, Carol, and Thomas Huckin. "News Value." In *Genre Knowledge in Disciplinary Communication*, 27–44. Mahwah, NJ: Erlbaum, 1995.

Berkove, Lawrence I. "Connecticut Yankee: Twain's Other Masterpiece." In *Making Mark Twain Work in the Classroom*, edited by James S. Leonard, 318. Durham, NC: Duke University Press, 1999.

Betts, John Richard. "P. T. Barnum and the Popularization of Natural History." *Journal of the History of Ideas* 20 (1959): 353–68.

Bickham, William Dennison. *A Buckeye in the Land of Gold: The Letters and Journals of William Dennison Bickham*. Edited by Randall E. Ham. Spokane, WA: Clark, 1996.

Blair, Walter, and Hamlin Hill. *America's Humor: From Poor Richard to Doonesbury*. New York: Oxford University Press, 1978.

Blutner, Reinhard, and Henk Zeevat, eds. *Optimality Theory and Pragmatics*. Palgrave Studies in Pragmatics, Languages, and Cognition. New York: Palgrave Macmillan, 2004.

Boese, Alexander. *The Museum of Hoaxes*, http://www.museumofhoaxes.com (accessed May 5, 2003).

———. *The Museum of Hoaxes: A Collection of Pranks, Stunts, Deceptions, and Other Wonderful Stories Contrived for the Public from the Middle Ages to the New Millennium*. New York: Dutton, 2002.

Booth, Wayne C. *The Rhetoric of Fiction*. Chicago: University of Chicago Press, 1961.

Brown, Lee Rust. *The Emerson Museum*. Cambridge, MA: Harvard University Press, 1997.

Bruce, Robert V. *The Launching of Modern American Science, 1846–1876*. First ed. New York: Knopf, 1987.

Bryant, John. "Poe's Ape of Unreason: Humor, Ritual, and Culture." *Nineteenth-Century Literature* 51 (1996): 16–52.

Burnam, Tom. "Mark Twain and the Paige Typesetter: A Background for Despair." *Western Humanities Review* 6 (1951): 29–36.

Campbell, Karlyn Kohrs. "Consciousness-Raising: Linking Theory, Criticism, and Practice." *Rhetoric Society Quarterly* (2002): 45–64.

Carter, Steve. "A Possible Source for 'The Facts in the Case of M. Valdemar.'" *Poe Studies* 12 (1979): 36.

Charney, Davida. "A Study in Rhetorical Reading: How Evolutionists Read 'The Spandrels of San Marco.'" In *Understanding Scientific Prose*, edited by Jack Selzer, 203–31. Madison: University of Wisconsin Press, 1993.

Charvat, William. *The Profession of Authorship in America, 1800–1870: The Papers of William Charvat*. Edited by Matthew J. Bruccoli. Columbus: Ohio State University Press, 1968.

Clemens, Samuel Langhorne. "3,000 Years among the Microbes." In *Which Was the Dream?* edited by John S. Tuckey, 433–553. Berkeley: University of California Press, 1968.

———. "The American Claimant." In *The American Claimant and Other Stories and Sketches*. New York: Harper & Brothers, 1923.

———. *The Autobiography of Mark Twain*. Edited by Charles Neider. New York: HarperCollins, 2000.

———. "Double-Barreled Detective Story." *Harper's New Monthly*, January 1902.

———. *Early Tales and Sketches*. Edited by Edgar Marquess Branch, Robert H. Hirst, and Harriet Elinor Smith. Vol. 1, 1851–1864, *Works of Mark Twain*. Berkeley: University of California Press, 1979.

———. *Mark Twain of the Enterprise: Newspaper Articles & Other Documents 1862–1864*. Edited by Henry Nash Smith. Berkeley: University of California Press, 1957.

———. *Mark Twain's Letters*. Edited by Edgar Marquess Branch, Michael B. Frank, and Kenneth M. Sanderson. Vol. 1, 1853–1866, *The Mark Twain Papers*. Berkeley: University of California Press, 1988.

———. "Memoranda: A Couple of Sad Experiences." *Galaxy* 9 (1870): 858–61.

———. "A Private History of a Campaign That Failed." In *The American Claimant and Other Stories and Sketches*. New York: Harper & Brothers, 1923.

———. *Roughing It*. Hartford, CT: American, 1891.

———. "To the Person Sitting in Darkness." In *Mark Twain's Weapons of Satire: Anti-Imperialist Writings on the Philippine-American War*, edited by Jim Zwick. Syracuse: Syracuse University Press, 1992.

Coleridge, Samuel Taylor. *Biographia Literaria*. Edited by J. Shawcross. Vol. 2. London: Oxford University Press, 1949.

Collins, Paul. *Banvard's Folly: Thirteen Tales of People Who Didn't Change the World*. New York: Picador USA, 2001.

Cook, James W. *The Arts of Deception: Playing with Fraud in the Age of Barnum*. Cambridge, MA: Harvard University Press, 2001.

Cooter, Roger, and Stephen Pumfrey. "Separate Spheres and Public Places: Reflections on the History of Science Popularization and Science in Popular Culture." *History of Science* 32 (1994): 237–67.

Cosmides, Leda, and John Tooby. "Consider the Source: The Evolution of Adaptations for Decoupling and Metarepresentation." In *Metarepresentation*, edited by Dan Sperber. New York: Oxford University Press, 2000.

Covici, Jr., Pascal. *Mark Twain's Humor: The Image of a World*. Dallas: Southern Methodist University Press, 1962.

Cox, James M. *Mark Twain: The Fate of Humor*. Princeton, NJ: Princeton University Press, 1966.

Cummings, Sherwood. *Mark Twain and Science: Adventures of a Mind*. Baton Rouge: Louisiana State University Press, 1988.

Dinius, Marcy J. "'Hans Phaall,' the Great Moon Hoax, and Popular Science in the Antebellum American Imaginary." Paper presented at the Second Annual International Poe Conference, Baltimore, October 3–6, 2002.

Drury, Wells. *An Editor on the Comstock Lode*. Reno: University of Nevada Press, 1984.

Dwyer, Richard A., and Richard E. Lingenfelter. *Dan De Quille, the Washoe Giant: A Biography and Anthology*, Western Literature Series. Reno: University of Nevada Press, 1990.

Eberly, Rosa A. *Citizen Critics: Literary Public Spheres*. Urbana: University of Illinois Press, 2000.

Eco, Umberto. *Serendipities: Language and Lunacy*. New York: Columbia University Press, 1998.

Edgar Allan Poe Society. "Poe's Letters." *Edgar Allan Poe Society*,. http//www.eapoe.org/misc/letters/t4604000.htm (accessed January 10, 2001).

Eisenstein, Elizabeth. *The Printing Press as an Agent of Change: Communications and Cultural Transformations in Early Modern Europe*. New York: Cambridge University Press, 1979.

Elmer, Jonathan. *Reading at the Social Limit: Affect, Mass Culture, and Edgar Allan Poe*. Stanford: Stanford University Press, 1995.

Emerson, Everett H. *Mark Twain: A Literary Life*. Philadelphia: University of Pennsylvania Press, 2000.

Emerson, Ralph Waldo. *Early Poems of Ralph Waldo Emerson*. New York: Crowell, 1899.

Fahnestock, Jeanne. "Accomodating Science: The Rhetorical Life of Scientific Facts." *Written Communication* 3, no. 3 (1986): 275–96.

———. "Rhetoric of Science: Enriching the Discipline." *Technical Communication Quarterly* 14, no. 3 (2005): 277–86.

Fahnestock, Jeanne, and Marie Secor. "Rhetorical Analysis." In *Discourse Studies in Composition*, edited by Ellen Barton and Gail Stygall, 177–200. New Jersey: Hampton, 2002.

———. "The Stases in Scientific and Literary Argument." *Written Communication* 5 (1998): 427–43.

Falk, Doris V. "Thomas Low Nichols, Poe, and the "Balloon Hoax."" *Poe Studies* 5, no. 2 (1972): 48–49.

Fedler, Fred. *Media Hoaxes*. Ames: Iowa State University Press, 1989.

Ferguson, DeLancey. "The Petrified Truth." *Colophon II* (n.s.) 2 (1937): 189–196.

Fish, Stanley. "Professor Sokal's Bad Joke." In *The Sokal Hoax: The Sham That Shook the Academy*, edited by Editors of *Lingua Franca*, 81–84. Lincoln: University of Nebraska Press, 2000.

Fisher IV, Benjamin Franklin. "Poe's 'Tarr and Fether': Hoaxing in the Blackwood Mode." In *The Naiad Voice: Essays on Poe's Satiric Hoaxing*, edited by Dennis W. Eddings, 136–47. Port Washington, NY: Associated Faculty, 1983.

Fishkin, Shelley Fisher. *Lighting out for the Territory: Reflections on Mark Twain and American Culture*. New York: Oxford University Press, 1996.

Fleck, Ludwik. *Genesis and Development of a Scientific Fact*. Translated by Fred Bradley and Thaddeus J. Trenn. Edited by Thaddeus J. Trenn and Robert K. Merton. Chicago: University of Chicago Press, 1981.

Franca. Editors of *Lingua*, ed. *The Sokal Hoax: The Sham That Shook the Academy*. Lincoln: University of Nebraska Press, 2000.

Franklin, H. Bruce. *Future Perfect: American Science Fiction of the Nineteenth Century*. Revised ed. New York: Oxford University Press, 1978.

———. "Traveling in Time with Mark Twain." In *American Literature and Science*, edited by Robert J. Scholnick, 157–71. Lexington: University Press of Kentucky, 1992.

Fusco, Richard. "Poe's Revisions of 'The Mystery of Marie Roget': A Hoax?" In *Poe at Work: Seven Textual Studies*, edited by Benjamin Franklin Fisher IV, 91–99. Baltimore: Poe Society, 1978.

Gervais, Bertrand. "Reading Tensions: Of Sterne, Klee, and the Secret Police." *New Literary History* 26 (1995): 855–84.

Goodwin, C. C. "Dan and His Quills." *Territorial Enterprise*, February 3 1876, 2.

———. "Dan De Quille." In *As I Remember Them*, 213–17. Salt Lake City, UT: Salt Lake Commercial Club, 1913.

Graff, Gerald. *Professing Literature: An Institutional History*. Chicago: University of Chicago Press, 1987.

Grice, H. Paul. "Logic and Conversation." In *Syntax and Semantics*, edited by Peter Cole and J. Morgan, 41–58. New York: Academic, 1975.

Griffin, Dustin H. *Satire: A Critical Reintroduction*. Lexington: University Press of Kentucky, 1994.

Gross, Paul, and Norman Levitt. *Higher Superstition: The Academic Left and Its Quarrels with Science*. Baltimore: Johns Hopkins University Press, 1994.

Guthrie, James R. "Broken Codes, Broken Seals, and Stolen Poems in 'The Purloined Letter.'" *Edgar Allan Poe Review* 3, no. 2 (2002): 92–102.

Hall, Bruce. "Grice, Discourse Representation, and Optimal Intonation." *Papers from the Regional Meetings, Chicago Linguistic Society* 2 (1998): 63–78.

Hall, David D. "Readers and Reading in America: Historical and Critical Perspectives." *Proceedings of the American Antiquarian Society* 103, no. 2 (1993): 337–57.

Harding, Sandra, ed. *Racial Economy of Science*. Boulder, CO NetLibrary, Inc., 1999.

Harris, Neil. *Humbug: The Art of P. T. Barnum*. Boston: Little, Brown, 1973.

Hay, Carolyn D. "A History of Science Writing in the United States and of the National Association of Science Writers." Dissertation, Northwestern University, 1970.

Henkel, Jacqueline. *The Language of Criticism: Linguistic Models and Literary Theory*. Ithaca, NY: Cornell University Press, 1996.

———. "Linguistic Models and Recent Criticism: Transformational-Generative Grammar as Literary Metaphor." *PMLA* 105, no. 3 (1990): 448–63.

Herschel Family Papers. Harry Ransom Center, The University of Texas at Austin.

Hoffman, Andrew Jay. *Inventing Mark Twain: The Lives of Samuel Langhorne Clemens*. New York: Morrow, 1997.

Hoffman, Daniel. *Poe Poe Poe Poe Poe Poe Poe.* First ed. Garden City, NY: Doubleday, 1972.

Hume, Beverly A. "Twain's Satire on Scientists: Three Thousand Years among the Microbes." *Essays in Arts and Sciences* 16 (1997): 71–84.

Katz, Stephen B. "Narration, Technical Communication, and Culture: *The Soul of a New Machine* as Narrative Romance." In *Constructing Rhetorical Education.*, edited by Marie Secor and Davida Charney, 382–402. Carbondale: Southern Illinois University, 1992.

Kaufer, David S., and Kathleen M. Carley. *Communication at a Distance.* Hillsdale, NJ: Erlbaum, 1993.

Kaufer, David, Suguru Ishizaki, Brian Butler, and Jeff Collins. *The Power of Words: Unveiling the Speaker and Writer's Hidden Craft.* Mahwah, NJ: Erlbaum, 2004.

Kennedy, George. "Literary Rhetoric." In *Classical Rhetoric and Its Christian and Secular Tradition from Ancient to Modern Times,* 108–19. Chapel Hill: University of North Carolina Press, 1980.

Kenner, Hugh. *The Counterfeiters: An Historical Comedy.* Reprint ed. Baltimore: Johns Hopkins University Press, 1985. Reprint, 1985.

Ketterer, David. "The 'science Fiction' of Mark Twain." *Mosaic—A Journal for the Interdisciplinary Study of Literature* 16, no. 4 (1983): 59–82.

Kintsch, Walter. *Comprehension: A Paradigm for Cognition.* New York: Cambridge University Press, 1998.

Lagemann, Ellen Condliffe. "Education to 1877." In *The Reader's Companion to American History,* edited by Eric Foner and John A. Garraty. Boston: Houghton Mifflin, 1991.

Lascarides, Alex, and Nicholas Asher. "Temporal Interpretation, Discourse Relations, and Commonsense Entailment." *Linguistics and Philosophy* 16 (1993): 437–93.

Latour, Bruno. "Is There Science after the Cold War?" In *The Sokal Hoax: The Sham That Shook the Academy,* edited by editors of *Lingua Franca,* 124–26. Lincoln: University of Nebraska Press, 2000.

———. *We Have Never Been Modern.* Translated by Catherine Porter. Cambridge, MA: Harvard University Press, 1993.

Lauber, John. *The Making of Mark Twain: A Biography.* New York: American Heritage, 1985.

Lee, Judith Yaross. "Fossil Feuds: Popular Science and the Rhetoric of Vernacular Humor." *Essays in Arts and Sciences* 23 (1994): 1–20.

———. "(Pseudo-) Scientific Humor." In *American Literature and Science,* edited by Robert J. Scholnick, 128–56. Lexington: University Press of Kentucky, 1992.

Lehrer, Keith, and Thomas D. Paxson Jr. "Knowledge, Undefeated Justified True Belief." In *Essays on Knowledge and Justification,* edited by George S. Pappas and Marshall Swain. Ithaca, NY: Cornell University Press, 1978.

Limon, John. *The Place of Fiction in the Time of Science*. New York: Cambridge University Press, 1990.

Ljungquist, Kent P. "'Valdemar'and the 'Frogpondians': The Aftermath of Poe's Boston Lyceum Appearance." In *Emersonian Circles: Essays in Honor of Joel Myerson*, edited by Wesley T. Mott, 181–206. Rochester, NY: University of Rochester Press, 1997.

Locke, Richard Adams. *The Celebrated "Moon Story," Its Origin and Incidents; / with a Memoir of the Author, and an Appendix, Containing, I. An Authentic Description of the Moon; Ii. A New Theory of the Lunar Surface, in Relation to That of the Earth*. Edited by William N. Griggs. New York: Bunnell and Price, 1852.

———. *The Moon Hoax; or, a Discovery That That Moon Has a Vast Population of Human Beings, The Gregg Press Science Fiction Series*. Boston: Gregg, 1975.

Loomis, C. Grant. "The Tall Tales of Dan De Quille." *California Folklore Quarterly* 5, no. 1 (1946): 26–71.

Mabbott, Thomas O., ed. *Doings of Gotham in a Series of Letters by Edgar Allan Poe as Described to the Editor of the Columbia Spy; Together with Various Editorial Comments and Criticisms, by Poe*. Pottsville, PA: Spannuth, 1929.

Machor, James L., ed. *Readers in History: Nineteenth Century American Literature and the Contexts of Response*. Baltimore: Johns Hopkins University Press, 1993.

Mailloux, Steven. *Rhetorical Power*. Ithaca: Cornell University Press, 1989.

Marks, Paula Mitchell. *Precious Dust: The American Gold Rush Era: 1848–1900*. New York: Morrow, 1994.

Martin, R. "Resurfacing Socialism: Resisting the Appeals of Tribalism and Localism." *Social Text* 44, Autumn–Winter (1995): 97–118.

Martineau, Harriet. *Retrospect of Western Travel*. New York, 1838.

Marx, Leo. *The Machine in the Garden: Technology and the Pastoral Ideal in America*. New York: Oxford University Press, 1964. Reprint, 2000.

Matthew, Marie-Louise Nickerson. "Forms of Hoax in the Tales of Edgar Allan Poe." Dissertation Columbia University, 1975.

Miller, Carolyn R. "Genre as Social Action." *Quarterly Journal of Speech* 70, no. 2 (1984): 151–67.

Miller, Howard S. "The Political Economy of Science." In *Nineteenth-Century American Science: A Reappraisal*, edited by George H. Daniels, 95–112. Evanston, IL: Northwestern University Press, 1972.

Mott, Frank Luther. *American Journalism: A History*. New York: Macmillan, 1962.

———. *A History of American Magazines: 1741–1850*. 5 vols. Vol. 1. Cambridge, MA: Harvard University Press, 1968.

———. *A History of American Magazines: 1850–1865*. 5 vols. Vol. 2. Cambridge, MA: Harvard University Press, 1968.

Muir, John. "Nevada's Dead Towns." In *Steep Trails*, edited by William Frederic Badè, 195–203. Boston: Houghton Mifflin, 1918.

Nelkin, Dorothy. *Selling Science: How the Press Covers Science and Technology*. revised ed. New York: Freeman, 1995.

O'Brien, Fitz-James. "How I Overcame My Gravity." *Harper's New Monthly Magazine* 28, no. 163 (1863): 779–82.

Ong, Walter J. *Orality and Literacy: The Technologizing of the Word*. New York: Methuen, 1982.

Orvell, Miles. *The Real Thing: Imitation and Authenticity in American Culture 1880–1940*. Chapel Hill: University of North Carolina Press, 1989.

Osborne, Peter. "Friendly Fire: The Hoaxing of *Social Text*." In *The Sokal Hoax: The Sham That Shook the Academy*, edited by editors of *Lingua Franca*, 195–99. Lincoln: University of Nebraska Press, 2000.

Paine, Albert Bigelow. *Mark Twain: A Biography*. New York: Harper & Brothers, 1912.

Paul, Danette, Davida Charney, and Aimee Kendall. "Moving beyond the Moment: Reception Studies in the Rhetoric of Science." *Journal of Business and Technical Communication* 15, no. 3 (2001): 372–99.

Perlman, David. "Science and the Mass Media." In *Science and Its Public: The Changing Relationship*, edited by Holton G. and W. Blanpied, 245–60. Dordrecht and Boston: Dorrecht, 1976.

Poe, Edgar Allan. "Astounding News by Express, via Norfolk!—The Atlantic Crossed in Three Days! Signal Triumph of Mr. Monck Mason's Flying Machine!" Extra, New York *Sun*, 13 April 1844.

———. *Eureka: A Prose Poem*. New York: Prometheus Books, 1997.

———. "The Facts in the Case of M. Valdemar." *The American Review: A Whig Journal*, December 1845, 561–65.

———. "Fifty Suggestions." In *Edgar Allan Poe: Essays and Reviews*, edited by G. R. Thompson, 1297–308. New York: Literary Classics of the United States, 1984.

———. "Hans Phaall—A Tale." *Southern Literary Messenger*, June 1835, 565–80.

———. *Letters of Edgar Allan Poe*. Edited by John Ward Ostrom. 2 vols. Cambridge, MA: Harvard University Press, 1948.

———. "Letter to B———." In *Edgar Allan Poe: Essays and Reviews*, edited by G. R. Thompson, 5–12. New York: Literary Classics of the United States, 1984.

———. "Literati—Laughton Osborn." Edgar Allan Poe Collection, Harry Ransom Center, University of Texas at Austin.

———. "The Literati of New York City: Richard Adams Locke." In *Edgar Allan Poe: Essays and Reviews*, ed. G. R. Thompson. E. A. Poe Society, http://www.eapoe.org/works/misc/litratb6.htm (accessed 4 January, 2001).

———. "Ms Found in a Bottle." In *Complete Stories and Poems of Edgar Allan Poe*, 148–55. New York: Doubleday, 1966.

———. "Murders in the Rue Morgue." In *Complete Stories and Poems of Edgar Allan Poe*, 2–26. New York: Doubleday, 1938.

———. "Note to 'Hans Phaall.'" In *The Moon Hoax*, edited by Ormond Seavey, 69–74. Boston: Gregg, 1975.

———. "A Predicament." In *The Complete Tales and Poems of Edgar Allan Poe*. New York: Modern Library, 1938.

———. "Sonnet—To Science." In *The Raven and Other Poems*, 55. New York: Wiley and Putnam, 1845.

———. "Von Kempelen and His Discovery." In *The Complete Tales and Poems of Edgar Allan Poe*. New York: Modern Library, 1938.

Pollin, Burton R. "Poe's 'Von Kempelen and His Discovery': Sources and Significance." *Etudes Anglaises* 20 (1967): 12–23.

Porter, Theodore M. *Trust in Numbers: The Pursuit of Objectivity in Science and Public Life*. Princeton, New Jersey: Princeton University Press, 1995.

Prince, Alan, and Paul Smolensky. *Optimality Theory: Constraint Interaction in Generative Grammar*. Ms. ed. New Brunswick and Boulder: Rutgers University and University of Colorado, 1993.

Quinn, Arthur Hobson. *Edgar Allan Poe: A Critical Biography*. New York: Appleton-Century-Crofts, 1941.

Robbins, Bruce, and Andrew Ross. "Response: Mystery Science Theater." In *The Sokal Hoax: The Sham That Shook the Academy*, edited by editors of *Lingua Franca*, 54–58. Lincoln: University of Nebraska Press, 2000.

Robins, R. H. *A Short History of Linguistics*. Edited by R. H. Robins, Geoffrey Horrocks, and David Denison. 4th ed, *Longman Linguistics Library*. London: Longman, 1997.

Roggenkamp, Karen S. H. 2001. *The Sun, the Moon, and Two Balloons: Edgar Allan Poe, Literary Hoaxes, and Penny-Press Journalism*, http://roggenkamp.com/poe.html (accessed 11 March, 2002).

Ross, Joan Belcourt. "Mark Twain and the Hoax." Dissertation, Purdue, 1974.

Rourke, Constance. *American Humor*. New York: Harcourt Brace, 1931.

Russett, Cynthia E. *Darwin in America*. San Francisco: Freeman, 1976.

Sampson, Geoffrey. *Schools of Linguistics*. Stanford: Stanford University Press, 1980.

Schauber, Ellen, and Ellen Spolsky. *The Bounds of Interpretation*. Stanford: Stanford University Press, 1986.

Seltzer, Mark. *Bodies and Machines*. New York: Routledge, 1992.

Shapin, Steven. "Pump and Circumstance: Robet Boyle's Literary Technology." *Social Studies of Science* 14 (1984): 481–520.

———. "Science and the Public." In *The Companion to the History of Modern Science*, edited by R. C. Olby, G. N. Cantor, J. R. R. Christie, and M. J. S. Hodge, 990–1006. New York: Routledge, 1990.

Shepperson, Wilbur S. *Mirage-Land: Images of Nevada*. Reno: University of Nevada Press, 1992.

Sicherman, Barbara. "Sense and Sensibility: A Case Study of Women's Reading in Late-Victorian America." In *Reading in America: Literature and Social History*, edited by Cathy N. Davidson. Baltimore: Johns Hopkins University Press, 1989.

Snow, C. P. *The Two Cultures: And a Second Look*. New York: New American Library, 1963.

Sokal, Alan. "A Physicist Experiments with Cultural Studies." *Lingua Franca* May/June (1996): 62–64.

———. "A Plea for Reason, Evidence, and Logic." Paper presented at the Forum on the Sokal Hoax, New York University, October 30 1996, http://www.physics.nyu.edu/~as2/nyu_forum.html (accessed September 15, 2003).

———. "Reply to Michael Bérubé." In *The Sokal Hoax: The Sham That Shook the Academy*, edited by editors of *Lingua Franca*, 143–45. Lincoln: University of Nebraska Press, 2000.

———. "Revelation: A Physicist Experiments with Cultural Studies." In *The Sokal Hoax: The Sham That Shook the Academy*, edited by editors of *Lingua Franca*, 49–53. Lincoln: University of Nebraska Press, 2000.

———. "Transgressing the Boundaries: Towards a Transformative Hermeneutics of Quantum Gravity." *Social Text* 46/47 (1996): 217–52.

Sokal, Alan, and Jean Bricmont. *Intellectual Impostures: Postmodern Philosophers' Abuse of Science*. London: Profile, 1999.

Sperber, Dan, and Deirdre Wilson. *Relevance: Communication and Cognition*. Cambridge, MA: Harvard University Press, 1986.

Stoehr, Taylor. *Hawthorne's Mad Scientists: Pseudoscience and Social Science in Nineteenth-Century Life and Letters*. Hamden, CT: Archon Books, 1978.

Stuart, Granville. *The Montana Frontier 1852–1864*. Lincoln: University of Nebraska Press, 1925. Reprint, 1977.

Swales, John. "Research Articles in English." In *Genre Analysis: English in Academic and Research Settings*, 110–76. Cambridge, UK: Cambridge University Press, 1990.

Swift, Jonathan. "A Modest Proposal." In *The Norton Anthology of English Literature*, vol. 1, 5th ed., edited by M. H. Albrams, 2181-87. New York: W. W. Norton & Co., 1986.

Swirski, Peter. *Between Literature and Science: Poe, Lem, and Explorations in Aesthetics, Cognitive Science, and Literary Knowledge*. Montreal: McGill-Queen's University Press, 2000.

Tebeaux, Elizabeth. *The Emergence of a Tradition: Technical Writing in the English Renaissance, 1475–1640*, Baywood's Technical Communications Series. Amityville, NY: Baywood, 1997.

Thomas, Dwight, and David K. Jackson. *The Poe Log: A Documentary Life of Edgar Allan Poe 1809–1859*. Boston, MA: Hall, 1987.

Thompson, G.R. "Is Poe's "A Tale of the Ragged Mountains" a Hoax?" *Studies in Short Fiction* 6 (1969): 454–60.

Tichi, Cecilia. *Shifting Gears: Technology, Literacy, and Culture in Modernist America.* Chapel Hill: University of North Carolina Press, 1987.

Tompkins, Jane P. "The Reader in History." In *Reader-Response Criticism: From Formalism to Post-Structuralism*, edited by Jane P. Tompkins, 201–32. Baltimore: Johns Hopkins University Press, 1980.

Tompkins, Jane. *Sensational Designs: The Cultural Work of American Fiction, 1790–1860.* New York: Oxford University Press, 1985.

Tyson, James L. *Diary of a Physician in California.* Oakland: Biobooks, 1955.

Van Dijk, Teun. "News Schemata." In *Studying Writing: Linguistic Approaches*, edited by S. Greenbaum and Cooper, 155–86. Beverly Hills, CA: Sage Books, 1986.

Vigliotta, Anthony. "U.S. Trade Dollars 1873–1885." In *FUN-Topics* 44:3, http://www.fun topics.com/fun_topics_v44n3_vigliotta.html (accessed 30 September 2002).

Voltaire. *Candide.* Translated by Shane Weller. New York: Dover, 1993.

Wadlington, Warwick. *The Confidence Game in American Literature.* Princeton: Princeton University Press, 1975.

Waggoner, Hyatt Howe. "Science in the Thought of Mark Twain." *American Literature* 8, no. 4 (1937): 357–70, 445–47.

Welch, Kathleen. *Electric Rhetoric.* Cambridge, MA: MIT Press, 1999.

Welsh, Susan Booker. "Edgar Allan Poe and the Rhetoric of Science." Dissertation, Drew University, 1986.

Wilson, James D. "'The Monumental Sarcasm of the Ages': Science and Pseudoscience in the Thought of Mark Twain." *South Atlantic Bulletin: A Quarterly Journal Devoted to Research and Teaching in the Modern Languages and Literatures* 40, no. 2 (1975): 72–82.

Winsor, D. A. "Communication Failures Contributing to the Challenger Accident: An Example for Technical Communicators." *IEEE Transactions on Professional Communication* 31, no. 3 (1988): 101–07.

Wright, William. letter, 23 August 1874.

———. *The Big Bonanza.* Edited by Oscar Lewis. New York: Apollo Editions, 1947.

———. Dan De Quille Papers. BHNC MSS P-G 246, The Bancroft Library, University of California, Berkeley.

———. *Dives and Lazarus.* Edited by Lawrence I. Berkove. Ann Arbor, MI: Ardis, 1988.

———. "Mystery of the Savage Sump." *Territorial Enterprise*, 19 February 1876.

———. "Quille Drops—Those Pre-Historic Tracks." *Territorial Enterprise*, 23 November 1885, 1.

———. *Reporting with Mark Twain.* San Francisco: California, 1893.

———. "Sad Fate of an Inventor." *Territorial Enterprise*, 25 July 1874.

———. "Traveling Stones." *Territorial Enterprise*, 26 October 1867, 3.

———. "The Wonder of the Age: A Silver Man." *Golden Era*, 5 February 1865, 3–4.

Zwaan, Rolf. "Effect of Genre Expectations on Text Comprehension." *Journal of Experimental Psychology: Learning, Memory, and Cognition* 20, no. 4 (1994): 920–33.

Zwick, Jim. 2002. "Mark Twain and Imperialism" in *Anti-Imperialism in the United States, 1898–1935*, http://www.boondocksnet.com/ai/twain/index.html (accessed 18 August, 2002).

Index

Adams, Henry, 130–31, 221
American literature, nineteenth-century, 9–11, 28–29, 51–225
Andresen, Julie Tetel, 35, 38, 39
Aronowitz, Stanley, 240n28
Asher, Nicholas, 83n67
assumptions, readers' relating to science news, 6, 33, 222; *see also* expectations, reader
Atkinson, Dwight, 7
Austin, J. L., 37, 40

Bacon, Francis, 58, 115; Baconian induction, 60, 115–16
Baird, Spencer, 55, 202, 204, 219
Balloon-Hoax. *See under* Poe, Edgar Allan
Barbosa, Pilar, 45
Barnes, Barry, 224
Barnum, P. T., 25, 55, 182, 195; Feejee Mermaid and, 25, 55
Barrett, Elizabeth B., 99, 101–2
Barritt, C. L., 58
Barthes, Roland, 9, 40, 52–53, 67, 98, 101–2
Bazerman, Charles, 5, 9, 42, 73, 76, 80, 135
belief, 54, 83, 112–13, 214 (*see also* epistemology; world views of readers); codependence of doubt and, 3, 187–88, 205, 209; doxastic systems of, 83n67
Benton, Richard, 110
Berkenkotter, Carol, 7, 9, 73
Berkove, Lawrence, 10, 158, 160, 174, 176, 207–8, 209–10
Bickham, William Dennison, 138
Blair, Walter, 125, 214n4
Blutner, Reinhard, 45
Boese, Alexander, 2, 107
Borges, Jorge Luis, 45–47
Booth, Wayne, 45
Branch, Edgar Marquess, 134
Bricmont, Jean, 233
Brown, Lee Rust, 59
Brown, Penelope, 167n111
Bruce, Robert V., 1, 76, 130
Bryant, John, 111
Burke, Kenneth, 31
Burnam, Tom, 159

Callon, Michel, 5
Campbell, Karlyn Kohrs, 214
Carley, Kathleen, 7, 19, 30, 205, 239n25
Charney, Davida, 7, 9, 135, 235n18
Charvat, William, 93
Chomsky, Noam, 37
Clemens, Orion, 122, 123–24, 129, 142, 221
Coleridge, Samuel Taylor, 58, 112
Cooter, Roger, 5, 31n35
Cope, Edward Drinker, 203–5, 219
counterfeit, 3
Covici, Pascal, 121, 125, 143–44, 158–59, 170
Cox, James, 167–68, 172

Culler, Jonathan, 9
cultural studies, 228–29, 234
culture: Jacksonian, 27–28, 31–32, 51, 55, 94, 126, 138, 178, 205; nineteenth-century literary, 28–29; nineteenth-century scientific, 27–28, 54–56, 98 (*see also* science: in antebellum America); two-cultures controversy, 30, 228
Cummings, Sherwood, 159

Darwin, Charles, 131–32
Davy, Sir Humphrey, 58, 102, 104–5
De Quille, Dan (William Wright), 1, 10–11, 16, 126, 138, 142, 145–46, 173–225, *181*; *Big Bonanza*, 11, 177, 194; *Dives and Lazarus*, 174, 210–11; Eyeless Fish hoax, 198–202, 205; and free silver, 200, 210; motivations for hoaxes, 173, 206–11, 216; previous work on the hoaxes of, 179, 207–8; and rhetoric, 173–76; and science, 176–78; Silver Man hoax, 176, 178, 179–88, 202; Solar Armor hoax, 1, 188–94, 202; technical reporting of, 174, 176–78, 207; Traveling Stones hoax, 194–98; and the West, 173, 187, 204, 206–11
de Saussure, Ferdinand, 37, 39
Defoe, Daniel, 19n9, 23, 103–4
Derrida, Jacques, 37–38, 40
Desmond, Adrian, 5
Drury, Wells, 177, 179, 195, 202, 207–8
Dwyer, Richard, 11, 173–74, 195

Eddings, Dennis, 110
Eisenstein, Elizabeth, 224
elenchus. *See* Socratic *elenchus*
Elmer, Jonathan, 55, 117–18
Emerson, Ralph Waldo, 29, 221
Eberly, Rosa, 42
Eco, Umberto, 197–98
Enterprise. See *Virginia City Territorial Enterprise*
epistemology: belief/doubt and, 3, 198; of the frontier, 124–25, 127, 187, 189, 209, 211, 215; literary, 2, 228–29; scientific, 4, 114–17, 227; social, 2, 4, 117–19, 164, 198, 208, 222, 227
ethnoscience: in antebellum America, 27–31, 54–56, 58; definition of, 4; in postbellum America, 129–32, 205
ethos, 93–94, 117–19, 168, 172, 177–78, 180, 186, 205, 207–8, 233–35, 237–38
Evening Bulletin. See *San Francisco Evening Bulletin*
expectations, reader: definition of, 10, 67, 83, 155–56; ranking, 85–89, 106–7, 183, 113–14, 134, 136, 146, 154–57 (*see also* filter, common reading); recuperating, 66–85, 133: relating to cultural studies, 228–232; science newsreading, 6, 8, 41–43, 45–49, 134–39, 142–47, 154–57, 183–88, 190–193, 197–98, 200–1, 203–6, 214, 223–24
expertise, 93–94; v. *elite*, 31n35; notoriety and, 4, 94, 232–35

Fahnestock, Jeanne, 5, 8, 20, 76, 235
Fedler, Fred, 2
Ferguson, DeLancey, 141
filter, common reading, 10, 15, 47, 83, 98, 102, 113–14, 133, 154–57, 192, 205–6, 218–19, 238
Firth, J. R., 40
Fish, Stanley, 9, 48, 234
Fisher, Benjamin Franklin, 110
Fishkin, Shelley Fisher, 159
Fleck, Ludwik, 5
Floy, Michael, Jr., 86
Foucault, Michel, 227
Franklin, H. Bruce, 26, 107, 130, 163

Geertz, Clifford, 111
genres, 2, 4, 68, 107–10; scientific, 6–7, 73–81, 145, 148–53, 186
Gervais, Bertrand, 135–36
Golden Era. See *San Francisco Golden Era*
Golinski, Jan, 6

Goodwin, C. C., 174–75, 178, 190, 199, 202
Graff, Gerald, 35–36
Grass Valley Union, 200–1
Great Wall of China hoax, 125n10
Grice, H. Paul, 11–12; maxims, 12, 42, 45, 47
Griffin, Dustin, 18–20
Griggs, William, 95, 168
Gross, Paul, 227

Hall, Bruce, 13
Harding, Sandra, 62n39
Harris, Neil, 182
Hay, Carolyn D., 57
Henkel, Jacqueline, 35, 37–39, 48–50
Henry, Joseph, 57
Herald. See *New York Herald*
Herschel, J. F. W., 42, 58, 63, *65*
Hill, Hamlin, 125, 214n4
Hine, Darlene Clark, 130n35
hoax (for hoaxes by individual authors, *see under* author's name): as rhetorical event, 2, 3–4; compared to fraud, 23–25; compared to parody, 7, 21–23, 234, 236; compared to satire, 18–21, 140–43, 234, 236; compared to science fiction, 26–27; compared to tall tale, 25–26 , 168–69, 207 (*see also* tall tale); as a computer virus, 235–40; definition of, 2, 4, 16–27, 49–50, 107, 118–19, 219–25; definition of scientific media hoax, 16, 219–25; folk definition of, 3, 24–25; Great Stock Exchange, 24–25; on the Internet, 227, 237–38; previous scholarship on, 2–4
Hoffman, Daniel, 104, 116–17
Huckin, Thomas, 7, 9, 73
Huizinga, Johan, 111
Humboldt, Alexander, 58
Hume, Beverly, 162
humor: and the Civil War, 125; laughter, 19, 132, 166–72, 215; practical joke, 125–27; and the West, 125 (*see also* tall tale)
Huxley, Thomas, 131

intention, authorial: fallacy of, 3; and hoaxing, 4, 49, 101, 154–55, 214–17
Iser, Wolfgang, 9

Jackson, David, 61
Jakobson, Roman, 40
journalism. *See under* media

kairos: definition of term, 3, 215; antebellum reading, 31–33; antebellum science/art, 4, 23, 27–31, 58
Katz, Steven, 31
Kaufer, David, 7, 19, 30, 205, 239n25
Kendall, Aimee, 7
Kennedy, George, 108
Kenner, Hugh, 3
Kintsch, Walter, 83

Lagemann, Ellen Condliffe, 74n58
Lascarides, Alex, 83n67
Latour, Bruno, 5, 222, 240n28
Law, John, 5
Lee, Judith Yaross, 54, 128–29, 176, 178
letteraturizzazione, 108
Levinson, S. C., 167n111
Levitt, Norman, 227
Limon, John, 114
Lingenfelter, Richard, 11, 173–74, 195
Lingua Franca, 227–28, 233–34
linguistics, 11–15; and literary criticism, 35–41; philology, 11, 35–36, 39–41. *See also* optimality theory; pragmatics
Ljungquist, Kent, 119
Locke, Richard Adams, 1, 51–52, 62–63, *64*, 115; Moon Hoax, 1, 13–15, 42, 55, 62–73, 81–90, 106, 168
Loomis, C. Grant, 178–79, 202, 207

Maelzel's chess-playing automaton, 3. *See also under* Poe, Edgar Allan
machines: hoaxes as, 16, 214, 236–37; in nineteenth century, 221–22
Machor, James, 9, 67
Mailloux, Steven, 24, 42
Marks, Paula Mitchell, 137–38
Martineau, Harriet, 65
Marx, Leo, 30, 221

Mason, Monck, 58, 90, 95
Matthew, Marie-Louise Nickerson, 110, 112–13
mechanism, 159–60; rhetorical, of hoaxes, 4, 94–97, 117, 125, 166–72, 190, 220–25
media: clipping in nineteenth-century news, 137, 190–93; hoaxes in, 3, 168, 237–38; mining news, in postbellum West, 137–39, 176–78; nineteenth-century, 165; popular scientific/technical, antebellum, 2, 31–32, 57–58, 73–81, 219; popular scientific/technical, postbellum, 145, 148–53, 186–87, 219; popular scientific/technical, twentieth century, 5, 31, 138n57; scientific, 28, 31–32; twentieth/twenty-first-century, 224–25, 237–38; women readers and nineteenth-century news, 62 (*see also* women: figured in popular science media); yellow journalism, 3
Melville, Herman, 59
mesmerism, 28, 56, 99–101
Miller, Carolyn R., 7
miners, gold rush, 137–39, 155–57, 182, *182*, 203–4, 209, 210–11 (*see also* media: mining news, in postbellum West)
Moon Hoax. *See under* Locke, Richard Adams
Mott, Frank Luther, 28, 74, 137
Muir, John, 213–14
Myers, Greg, 5, 6

Nelkin, Dorothy, 5, 31
New York Herald, 62–63, 72–73; ed. James Gordon Bennett, 62–64
New York Sun, 1, 41, 62–64, 66, 68, 89, 91–92, 190, 199
new-historicist criticism, 4, 9–11, 41–42
Nichols, Thomas Low, 92

O'Brien, Fitz-James, 27, 108–10
Ong, Walter, 224, 239n25
optimality theory, 8; in linguistics, 13, 43–45; applied to reading, 14–15, 45–50, 214 (*see also* appendix A)
Orvell, Miles, 220–21
Osborne, Peter, 237, 239
Osler, Dr. William (Dr. Egerton Yorrick Davis), 57

Paine, Alfred Bigelow, 127–28, 171
Paul, Danette, 7
Perlman, David, 138n57
Poe, Edgar Allan, 1, 15, 29, 51–119, 124–25, 135, 154, 168, 170, 173, 175, 178, 193, 207, 221; Balloon-Hoax, 33, 51, 90–97, 117, 144; burlesques, 22–23, 111; *Eureka*, 51, 58, 114–19, 215; Hans Phaall hoax, 22–23, 51, 60–62, 81, 94; M. Valdemar hoax, 51, 56, 67, 97–102, 216, 223; motivations for hoaxing, 52, 60, 81, 93, 116–19, 215–16; "MS Found in a Bottle," 59, 158; "Murders in the Rue Morgue," 111–13, 119; previous work on hoaxes of, 2, 110–13; relationship to Maelzel's chess-playing automaton, 3, 95, 103; relationship to readers, 116–19; and rhetoric, 52–54; and science, 54, 57–60, 116–17; and science fiction, 26–27; "Sonnet—To Science," 54, 60; Von Kempelen hoax, 51, 102–7
Pollin, Bernard, 104, 106
pragmatics, 2–4, 8, 45
preference rules, 42, 45
Prince, Alan, 13, 43
Pumfrey, Stephen, 5, 31n35

Quinn, Arthur Hobson, 105

Ramsay, Arch, 99–100
reactions, reader. *See* responses, reader
reader-response criticism, 4, 9–11, 37, 41
reception of hoaxes. *See* filter, common reading; expectations, reader
relevance theory, 42
responses, reader, 4, 8, 13–14, 41, 134–36, 139–41, 183. *See also* appendix B
Robins, R. H., 35
Roggenkamp, Karen, 118
Ross, Andrew, 240n28

Ross, Joan Belcourt, 157–58, 160–61, 170
Rourke, Constance, 170

Sampson, Geoffrey, 40
San Francisco Evening Bulletin, 132–36, 139, 143–44, 150
San Francisco Golden Era, 174–76, 179–80
Sarchett, Barry, 229n3
Schauber, Ellen, 42, 45
Scholes, Robert, 41
science: in antebellum America, 54–60; and the Civil War, 130–31; compared to engineering, 129; popularization of, 5–6, 211 (*see also* media: popular scientific/technical); in postbellum America, 150, 163, 166, 171, 186, 189; professionalization of, 28, 30, 31n35, 94; pseudoscience, 56; rhetoric of, 4–9, 150; social, 56; as social epistemology, 2, 94, 227; and technology studies, 3
Scientific American, 31, 74, 76, 148, 190
Searle, John, 37, 40
Seavey, Ormond, 26, 63
Secor, Marie, 20, 228, 235
Secord, James, 6
Shapin, Steven, 5, 80
Shepperson, Wilbur S., 23
Skaggs, Joey, 237n24
Smith, Henry Nash, 137
Smolensky, Paul, 13, 43
Snow, C. P., 30
Social Text, 227–40
Socratic *elenchus*, 20, 144, 169–70
Sokal, Alan, 227, 230–32, 234, 239; Sokal hoax, 3, 224, 227–240
Southern Literary Messenger, 60–61, 109
Sperber, Dan, 42
Spolsky, Ellen, 42, 45
stasis theory, 20, 239
Stoehr, Taylor, 59
Sun. See *New York Sun*
Swales, John, 7, 76
Swift, Jonathan, 17–21, 174; almanac hoax, 18; *A Modest Proposal*, 18–21
tall tale, 124–27, 167–68. *See also under* hoax
Symmes, John, 217

Tebeaux, Elizabeth, 80, 222
technical communication, 8–9, 10. *See also* De Quille: technical reporting; journalism: mining
Territorial Enterprise. See *Virginia City Territorial Enterprise*
Thomas, Dwight, 61
Thompson, G. R., 110
Tichi, Cecilia, 222
Tompkins, Jane, 9, 214n4
topoi, 6, 10, 76; pertaining to science news/hoaxes, 10, 42, 67–71
trade dollars, 200n51
Twain, Mark (Samuel Langhorne Clemens), 5, 10, 15–16, 121–73, 176, 207, 221; *The American Claimant*, 164–5; *Christian Science*, 170; *A Connecticut Yankee in King Arthur's Court*, 121, 158, 160; and Darwin, 128, 132–33; Empire City Massacre, 131, 139–46, 169, 175; "How the Animals of the Woods Sent Out a Scientific Expedition," 128, 132, 162; *Huckleberry Finn*, 24, 131, 158; motivations for hoaxing, 121–22, 139–40, 169, 215–16; "The Mysterious Stranger," 166, 172; The Petrified Man hoax, 1, 26, 33, 127, 132–47, 169–70, 190; previous work on hoaxes of, 157–61; and rhetoric, 122–27; *Roughing It*, 25, 125–26; and science, 121, 127–32, 159–66; social activism of, 166, 170–72; *3,000 Years among the Microbes*, 132, 161–63, 169–70; "To the Person Sitting in Darkness," 171
Tyson, James, 138

United States Exploring Expedition, 27, 55, 127; Wilkes expedition, 116

van Dijk, Teun, 83, 150
Vigliotta, Anthony, 200n51

Virginia City, NV, 1, 174–5, *185*, 191
Virginia City Territorial Enterprise, 1, 10, 123, 132, 136–37, 139, 146–48, 168, 173–76, 179, *184*, *185*, 188, 194–95, 200, 203, 206, 218
Voltaire (François-Marie Arouet), 124; *Candide*, 169

Wadlington, Warwick, 24, 172
Waggoner, Hyatt, 128–29, 159, 166
Welles, Orson and "War of the Worlds," 217
Welch, Kathleen, 224
Welsh, Susan Booker, 53
Whitney, William Dwight, 39–40
Wilson, Deirdre, 42
Wilson, James, 159, 166
Winsor, Dorothy, 8–9
women: and hoaxing, 214n4; figured in popular science media, 58n26
world views of readers, 2, 3, 83, 239–40

Zeevat, Henk, 45
Zwaan, Rolf, 113n122, 145
Zwick, Jim, 171